ADDITIONAL PRAISE FOR

The Organic Farmer's Business Handbook

"Generous, unique, and comprehensive—this book will help every farmer to better set goals, reach those goals, and improve their bottom line!"

—Ela Chapin, Program Director,
Vermont Farm Viability Enhancement Program

"How could anyone hope to produce a business guidebook useful to one of the most diverse, independent, even contrary groups of people on earth? This essential book is overdue! While the principles have been around for some time, this is the first they have been assimilated and tested so successfully."

—Ed Martsolf, Petit Jean Farm,
Morrilton, Arkansas

"Many people become organic farmers because they love growing crops and working the land, but truth be told that's often the easy part. Richard Wiswall's book provides practical, real-world guidance for dealing with the hard part: the business of farming."

—Vern Grubinger, Vegetable and Berry Specialist,
University of Vermont Extension

THE ORGANIC FARMER'S BUSINESS HANDBOOK

THE ORGANIC FARMER'S BUSINESS HANDBOOK

A Complete Guide to Managing Finances, Crops, and Staff – and Making a Profit

RICHARD WISWALL

Chelsea Green Publishing
White River Junction, Vermont

Project Manager: Patricia Stone
Developmental Editor: Benjamin Watson
Copy Editor: Laura Jorstad
Proofreader: Nancy Ringer
Indexer: Peggy Holloway, Holloway Indexing Services
Designer: Peter Holm, Sterling Hill Productions
Photographs and illustrations by Richard Wiswall unless otherwise noted

Printed in the United States of America
First printing, September 2009
10 9 8 7 6 5 4 3 2 1 09 10 11 12 13

Our Commitment to Green Publishing

Chelsea Green sees publishing as a tool for cultural change and ecological stewardship. We strive to align our book manufacturing practices with our editorial mission and to reduce the impact of our business enterprise in the environment. We print our books and catalogs on chlorine-free recycled paper, using vegetable-based inks whenever possible. This book may cost slightly more because we use recycled paper, and we hope you'll agree that it's worth it. Chelsea Green is a member of the Green Press Initiative (www.greenpressinitiative.org), a nonprofit coalition of publishers, manufacturers, and authors working to protect the world's endangered forests and conserve natural resources.

The Organic Farmer's Business Handbook was printed on Joy White, a 30-percent postconsumer recycled paper supplied by Thomson-Shore.

Chelsea Green Publishing Company
Post Office Box 428
White River Junction, VT 05001
(802) 295-6300
www.chelseagreen.com

Library of Congress Cataloging-in-Publication Data
Wiswall, Richard, 1957-
The organic farmer's business handbook : a complete guide to managing finances, crops, and staff; and making a profit / Richard Wiswall.
p. cm.
Includes index.
ISBN 978-1-60358-142-4
1. Organic farming. 2. Farm management. I. Title.

S605.5.W57 2009
631.5'84068--dc22

2009025039

Chelsea Green Publishing is committed to preserving ancient forests and natural resources. We elected to print this title on 30-percent postconsumer recycled paper, processed chlorine-free. As a result, for this printing, we have saved:

21 Trees (40' tall and 6-8" diameter)
9,431 Gallons of Wastewater
7 Million BTUs Total Energy
573 Pounds of Solid Waste
1,958 Pounds of Greenhouse Gases

Chelsea Green Publishing made this paper choice because we and our printer, Thomson-Shore, Inc., are members of the Green Press Initiative, a nonprofit program dedicated to supporting authors, publishers, and suppliers in their efforts to reduce their use of fiber obtained from endangered forests. For more information, visit: www.greenpressinitiative.org.

Environmental impact estimates were made using the Environmental Defense Paper Calculator. For more information visit: www.papercalculator.org.

For Sally,

Pete, Kuenzi, and Flint

CONTENTS

ABOUT THE COMPANION CD

The files on the companion CD are tools designed for use on your computer to save you time and money. The Farm Crew Job Description template is a Microsoft Word document and the Timesheet is a Microsoft Excel worksheet; both can be printed out to help with employee management on your farm. The other three files are Microsoft Excel workbooks, which contain self-calculating worksheets. You will probably want to copy the files from the CD onto your computer's hard drive and utilize them from there, retaining the CD as a backup. You cannot save changes to the copies of the files that are on the CD, but you will be able to save changes to copies that are saved to your hard drive.

The Vegetable Farm Crop Enterprise Budgets contain all the worksheets listed in the Appendix in the back of this book, plus two templates (one is partially filled in, one is not). Worksheets can be tailored to any farm's needs. Expense and sales names may be modified, along with their numbers, for different scenarios. The templates are easily duplicated. All the budgets are self-calculating. Budgets can also be used as projections for possible future changes or new crop trials. Use the budgets as a guide for your farm. Every farm has different systems, various costs associated with every component of the farm, and a unique blend of sales revenue. The best numbers are your own.

The Payroll Calculator simplifies payroll processing. It calculates employee gross wages, subtracts out Social Security, Medicare, federal and state withholding taxes to yield net pay, and tabulates employer tax liabilities by the month. The Payroll Calculator is designed for semimonthly paychecks for up to eight employees. All you need is a withholding tax table from your home-state tax department (available from their website or local office) and Publication 51 (Circular A) *Agricultural Employer's Tax Guide* from the IRS (available at your local IRS office or by mail).

Last, the Vermont Farm Viability Enhancement Program Farm Financials Workbook is a comprehensive financial-analysis tool used to create important financial documents such as Balance Sheets, Income Statements, and Cash Flow Projections. Additionally, a scorecard evaluates key financial indicators of your farm business.

The Farm Crew Job Description template and Timesheet are ready to print out for use on your farm, with only minor changes to reflect your farm. The other three files are for making your own calculations (and printing out as well). If you are new to Excel workbooks, begin by clicking on the tabs along the bottom of the screen to view the different worksheets. Enter your numbers into the relevant cells; the worksheet is set to calculate useful results based on the numbers you provide. The cells that contain formulas (subtotals, for example) are protected from accidental erasing. You can unprotect the formula cells if you wish. When in doubt, try the HELP button at the top of the screen. Numerous books, workshops, and online resources are available to learn more about working with Excel workbooks.

All of these files were created using Microsoft Office programs (Word and Excel). If you do not have Microsoft Office, you should still be able to use most or all of the files through compatible software packages, such as Apple iWork (for Apple computers only), Microsoft Works (for computers running Microsoft Windows only), or OpenOffice. OpenOffice is free to download at www.openoffice.org and is available for a variety of operating systems, including Microsoft Windows, Apple OS, and Linux. Neither I nor my publisher has thoroughly tested the Excel files in these alternative programs, so we can't guarantee that they will retain their full functionality, or function entirely properly, when opened in something other than Microsoft Office. In our limited testing, it appears that OpenOffice offers the closest approximation to using the files in their native Microsoft Office habitat. If you are using Microsoft Works, we expect that it should be okay for the Farm Crew Job Description, Timesheet, and Vegetable Farm Crop Enterprise Budgets, but we do *not* think that it will properly handle the Payroll Calculator or Vermont Farm Viability Enhancement Program Farm Financials Workbook, due to their use of more-complicated formulas in certain calculations.

ACKNOWLEDGMENTS

For great suggestions and editing help, thanks to Christa Alexander and Mark Fasching, Peter Griffin, Wendy Sue Harper, Laura Mahan, John and Joy Primmer, and John Wiswall. Thanks to Rowan Jacobsen and Laura Williams McCaffery for getting the ball rolling. The Financials Workbook, included on the accompanying CD, is courtesy of the Vermont Farm Viability Enhancement Program. Kudos to Chelsea Green Publishing, which walks the green talk and is a pleasure to work with. I'm indebted to Ed Martsolf for busting paradigms; his lessons germinated the seed for this book. And lastly, many, many thanks to Sally for her unwavering help and support.

– 1 –
True Sustainability

A few years ago at a New England Vegetable and Fruit Conference, I presented a talk on farm profitability with a fellow farmer. He opened the talk by saying, "Sometimes I think I should have listened to my parents and become a doctor or a lawyer—but you know, I don't think I could take the pay cut." Wow.

Here was a vegetable grower standing up in front of a room full of farmers and telling them he makes more money than doctors or lawyers. He was serious. Jolted as the audience was by that seismic statement, I knew I had a tough act to follow.

Farming conferences are terrific sources of information—seasoned farmers share their experience and knowledge, and agricultural professionals update attendees with the latest research and news. But the overwhelming majority of information at a conference focuses on aspects of production—how to grow crops, which seed varieties are hot, which tractors and tools increase efficiency, and pests and diseases to watch for. Very few presentations address the business side of farming.

Similarly, farming books almost all focus on production. Yet good production techniques alone will not make an organic farm sustainable. Most people go into organic farming with a love for the land and for growing food, and that love is essential to staying committed through the years of hard work. Too many farmers, however, never consider a farm's profit potential, or the various costs of production that ensure its financial health and longevity—and all too often they burn out because of it.

Organic farms comprise many different enterprises that get averaged out financially in a year-end profit or loss. A diversified organic vegetable farm may grow forty or more different crops, such as kale, broccoli, and sweet corn. Even a dairy farm with one product, milk, has different enterprises: milk cows, heifers, calves, silage, hay, and grain. Thank goodness for the IRS. Annual tax filing is often the only reason farmers look at their bottom line; without a Schedule F, the farm's current checkbook balance would be the only indicator of financial health.

Production techniques rarely limit a farm's success; rather it is the lack of *dependable profitable returns.* Farmers enjoy their work for lots of reasons: sowing seeds, working the soil, marveling at the plants that grow. Fundamental satisfaction comes from producing food, working outdoors, being your own boss, and working intimately with nature. No one's motivation to farm came from the desire to be better versed in IRS employee tax codes and workers' compensation laws, or to learn about pro forma balance sheets. Yet the farming and business worlds inevitably collide, and farmers are often uninformed about the business concepts and tools crucial to navigating forward effectively and profitably.

The information that follows draws on decades of personal farming experience and my thirst for smart and appropriate business tactics. I know firsthand the joys, frustrations, stresses, and challenges of starting and operating an organic farm. Contrary to what most people believe, a good living can be made on an organic farm, and what's required is *farming smarter, not harder.*

My goal is to highlight the necessary tools for successful and profitable farming for new and seasoned farmers alike. This first chapter starts with some "soft" business concepts, to lay the foundation for the practical step-by-step road to profitability.

The Mile-High Fence

Imagine a mile-high fence surrounding your farm or property. The fence is continuous along the outside perimeter of your land; it is open at the top so that sun and rain may enter, and it is porous for wind, birds, and insects to pass through. The air is naturally full of nitrogen, oxygen, and carbon dioxide, and the land is a living soil full of minerals, microbes, and organic matter. There are no breaks in the fence, except for one small gate. Your job as farmer is to monitor what goes in and out of that gate.

Most farms bring in lots of material like fuel, fertilizer, seed, and packaging; mix them up and change them a bit; then send them back out the gate. When you think about it, this business model isn't much different from that of a plastics factory. And yet our farms should not be places where petroleum-based inputs are turned into food. Our ultimate job as organic farmers is to use what is freely available to us in nature to generate *true* wealth. In this light, I see farming as one of the noblest endeavors: a real generator of healthy products using natural cycles.

The Mile-High Fence analogy is a novel way of looking at farming, placing the responsibility

Cows grazing. Credit: Vern Grubinger

for monitoring farm inputs and outputs on the farmer. As an organic farmer, what are some things that come in and go out of your farm gate?

To answer that question, I'll start with a simple and idealistic model of a dairy farm: Sun, rain, and atmospheric nitrogen, oxygen, and carbon dioxide photosynthesize in grass that is growing in the living soil. Cows feed on the grass, drink water (from the rain), and mature and rear their young. Cows are milked, the milk is exported through the farm gate, and money from the sale of this milk is brought back to the farm. Nutrients in the manure from the animals recycle within the farm system. Milk leaving through the farm gate is mostly water, and it is produced from grass grown with free sunshine, readily available elements in the environment, and soil nutrients, most of which are replenished with applications of manure. Given enough land, young stock are raised, and the process sustains itself indefinitely.

True sustainability is thus made possible by recycling what nutrients are readily available, and using rain and the energy from the sun.

So if we have all this free rain, solar energy,

nitrogen, carbon dioxide, oxygen, and microbially rich soil, shouldn't it be easy to make money farming? Sure, there are some obstacles, but a truly sustainable farm is based on these fundamental principles.

A number of years ago, the farm gate flow of my farm looked like this. Coming in the gate were:

- Borrowed money
- Seed
- Organic fertilizers
- Compost
- Fuel oil
- Laborers
- Organic pesticides
- Packaging—labels, bags, and boxes
- Greenhouse frames
- Greenhouse plastic
- Potting soil
- Plastic pots
- Electricity
- Telephone service
- Tractors, trucks, and other equipment
- Tools
- Parts for repairs
- Money from sales of farm products

Meanwhile, exiting the farm gate were:

- Produce raised on the farm
- Laborers returning home
- Trash, and payment for the landfill
- Loan payments
- Payments for seed, fertilizer, compost, fuel oil, laborers, and all the other purchased items listed above
- Payments for taxes, insurance, memberships, and trucking
- Payments for living expenses

This is a little more complicated than the simple example of sustainability in the dairy farm portrayed above, and a little more realistic for an organic vegetable farmer living in today's world. The job of the farmer standing by the one gate in the mile-high fence is to monitor what goes in and out. However, this is not to say that as many items as possible should be eliminated. Different farms have different inputs and outputs, and some farms have more than others. Importance needs to be placed on the relevance of each input and output to how it utilizes natural cycles.

Solar Dollars

Money is a medium of exchange. I use money to buy a chair, and I receive money when I sell a bag of carrots. I don't need to trade my carrots directly for the chair. Money is very handy, and it comes in various forms: coins, cash, checks, and credit cards, for example. But let's talk about the *origin* of money—what generated those dollars in the first place? I'm not talking about the printing press at the US Mint or the fractional reserve banking system, but rather a novel way of money classification.

A mentor of mine, Ed Martsolf of A Whole New Approach in Morrilton, Arkansas, taught concepts of Holistic Management,* which include some interesting ideas on money and on goal setting. Martsolf described money falling into three nontraditional but distinct types: mineral dollars, paper dollars, and solar dollars.

Mineral dollars are generated when products of value are mined or extracted and then sold. Gold, oil, coal, granite, and rock phosphate are some substances that generate mineral dollars. If I owned a quarry of granite, my sales would be in mineral dollars. The upside to mineral dollars is that the money from the granite is real and can provide a sizable income stream until the resource is used up. The downside is that its source is finite, and that eventual depletion of the resource will terminate the flow of mineral dollars. Mineral dollars are a one-way street.

Paper dollars are the most common of these three types of money. If I buy a tractor for $5,000 and immediately resell it for $5,500, I've made $500 in paper dollars. Paper dollars come from transactions. No real product is involved, just my time, and knowledge that an opportunity exists. With paper dollars, there is always a winner and a loser. One person profits at someone else's expense. Our financial institutions all deal in paper dollars: The stock market, banks, and businesses are all involved in buying and selling. There is no overall net gain in paper dollars unless the government or banks create more money.

Solar dollars are unlike mineral and paper dollars. Solar dollars generate true wealth. They are forever sustainable and transcend the winner–loser scenario. In the Mile-High Fence example, natural cycles use freely available components and the sun's energy to create a product of value—a product of solar dollars. The growing of plants and animals following basic natural cycles generates solar dollars. And while paper and mineral dollars may be used in conjunction with solar dollars, the focus on solar dollars is vital to any organic farm.

Paradigms

Learning about solar dollars caused a paradigm shift in my thinking. Paradigms are the set of rules and filters through which we view the world. Imagine a paradigm being the lens of a camera through which you see. Something within the camera's view is easily discernible, but anything outside it is not seen, even though it is really there.

If I show you a photograph of someone levitating, your immediate reaction will likely be that the picture has been touched up—that I've used some trick of photography. Why? Because the image doesn't reflect the commonly accepted paradigm of gravity.

The Swiss watch industry portrays a real-life paradigm shift to which we all can relate.

*For more information, visit www.holisticmanagement.org.

Futurist Joel Barker points out that in 1968, the Swiss were the experts in making watches, with 65 percent of the market share and 80 percent of the profits. Today the Swiss have less than 10 percent of the market share and less than 20 percent of the profits. Why such a steep decline?

In 1967, an innovative watchmaker in Switzerland realized that the qualities of quartz crystals could be utilized to keep precise time, and that this could be done very simply. The innovator proudly presented the new findings to the management teams of the established watch industry, expecting praise for the discovery. Instead, he was rebuffed. They looked at the simple quartz movement watch and dismissed it. It wasn't a watch: It lacked the conventional gears and springs.

Shortly thereafter, the quartz movement innovator attended the World Watch Conference to display his new idea. Representatives from Seiko and Texas Instruments took one look and snapped up the quartz movement idea. The rest is history. Quartz movement watches are now the industry standard.

Back to farming. If, after attending an organic farming conference, you notice that a lot of the participants drive away in luxury Jaguars, Ferraris, and Rolls-Royces, you instantly assume that the drivers are not real farmers making their living from the land. How could a farmer afford an expensive car with income derived solely from farming? These drivers must have family money in the bank, or they must pursue a different, more lucrative profession full-time. The operating paradigm is that farmers don't make much money and thus can't afford expensive cars.

What are some adjectives used to describe farmers? *Hardworking*? *Honest*? I grew up with two: *poor* and *dumb*. The *2000 American Heritage Dictionary* includes one outrageous definition of *farmer:* "A simple, unsophisticated person; a bumpkin." (I tried looking up the definition of a dictionary definition writer, but it was strangely absent.) The paradigm of the poor dumb farmer, unfortunately, is the dominant one in our culture. This very paradigm often becomes a self-fulfilling prophecy. If I'm not expected to make money at farming, I probably won't.

In 1993, I attended a class on Holistic Management, and the presenter leading the course in the front of the room said, "The biggest fallacy in farming is that there is no money in it."

Let me repeat that: *"The biggest fallacy in farming is that there is no money in it."*

I had to think about it for a second, and then I realized that what he had said made me angry. *What does he know about farming anyway? If he were a real farmer, he wouldn't be up there teaching.*

Picture in your mind a black-and-white photograph of a normal class, except for one guy whose face is beet red and steaming . . . that was me. I was doing everything I could do to discredit the speaker and what he said—because it didn't fit my paradigm.

Then he got me even more infuriated by saying, "You plan for profit before anything else."

What about droughts? Floods? *You plan for it.*

What about illness? *You plan for it.*

What about your barn burning down? *You plan for it.*

I almost walked out of the classroom right then and there. But he was right; I just couldn't see it yet. My paradigm was about to shift.

This "selective hearing" surrounding paradigms reminds me of a conversation I had with a vegetable farming neighbor of mine. I had been farming for five years at the time, and my neighbor enjoyed helping me out on occasion. He mentioned to me that he had visited with an established vegetable farmer in the region, who shared with my neighbor some financials. This established vegetable farmer said that he had made $100,000 the previous year. I asked, "Gross sales, right?" But the response was, "No, $100,000 net profit."

How could this be? I couldn't comprehend what was being said. In fact, I couldn't even recall that conversation until years later. It seemed unbelievable to me that a farmer could make that much money.

Years later, I co-presented the talk mentioned at the beginning of this chapter, in which my colleague said he sometimes regretted not becoming a doctor or lawyer like his parents wanted, but he couldn't handle the pay cut. From my vantage point in the room, I could see that, of all the people in the audience, only some actually heard *and absorbed* the statement. For the others, the words just passed over their heads like giant soap bubbles slowly floating along, bouncing off paradigm shields. Ears heard the words, yes, but minds didn't register them. The farmers with receptive minds, however, broke through the paradigm barrier and comprehended what was being said, and probably took some time to think about it.

Prosperous farmers do exist, but we need to hear more from them. More role models are needed to challenge the paradigm of the poor, dumb farmer. I've seen a number of farmers fail to recognize some highly profitable enterprises because ultimately they didn't think it was possible to make good money from farming. If a goose that laid golden eggs suddenly took up residence on their farm, the bird could be unappreciated and possibly even end up in the oven. One of the main goals of this book is to help create more prosperous, less stressed-out farmers. I'm not saying that farming is only about making money, but I am saying that a fair and decent return is both justified and necessary. I'm hoping the tools and information in these pages will help.

Goal Setting

Happiness is something we all strive for. I define *happiness,* in simple terms, as getting what you want. If this is true, it sounds easy enough. In reality, however, happiness can be quite elusive.

Getting what you want may not be that easy, mostly because we don't have a clear grasp on what exactly it is that we want. What we *truly want* is not a new tractor or late-model pickup truck. *Wants* need to be defined as deeper, value-based goals—things that are held very dear, such as family, creativity, leisure, health, or economic security. The new tractor or pickup truck is merely the outward manifestation of a desire for more economic security.

The exercises that follow are designed to uncover "wants" that are derived from your deeply held values. They are simple exercises to get your thoughts flowing. Your ideas and thoughts are yours and yours only. No one else needs to see your work, so be honest with yourself. The first two short exercises will take about ten minutes. The second part is an exercise that takes about half an hour. Block out the time it takes; don't answer the phone, and avoid other distractions. If you can't tackle it all now, come back to it later. Goal setting is a process, and its importance cannot be overstated. Clear goals are a fundamental building block for a successful farm.

Exercise One

If you had only $100 left, what would you do with it? How would you spend it? On what? There are no options for family or friends to lend you money or help you out . . . it is truly your last $100. This is just a hypothetical case, to get you thinking. Don't overanalyze the possible parameters—keep it simple. Take a piece of paper, put today's date on it, and write down your ideas.

Exercise Two

On the same piece of paper, write an obituary *for yourself*—a few sentences on what you think your obituary would look like (or what you'd like it to look like). What comes to mind? Again, no one will see this except you, so be honest.

So with your creative juices flowing, it is time for the final exercise. This will take about half an hour.

Personal Values Worksheet *(From Ed Martsolf, A Whole New Approach)*

Having contemplated some important values, fill out the following Personal Values Worksheet. This process is designed to further clarify values you hold dear.

Column A

Identify how well you feel you've *satisfied* each value. Use the following scale:

0 = not at all
1 = slightly
2 = some
3 = fairly well
4 = considerably well
5 = extremely well

Column B

Identify how you would feel if your current satisfaction of this value were significantly *reduced.* Use the following scale:

0 = not at all concerned
1 = slightly concerned

TABLE 1-1: Personal Values Worksheet

Personal Values	A	B	C	D
1. Accomplishment (achieving, mastery)				
2. Affection (close, intimate relationships)				
3. Collaboration (close working relationships)				
4. Creativity (imaginative self-expression)				
5. Economic security (prosperous, comfortable life)				
6. Exciting life (stimulating, challenging experiences)				
7. Family happiness (contentedness with loved ones)				
8. Freedom (independence and free choice)				
9. Inner harmony (serenity and peace)				
10. Order (stability and predictability)				
11. Personal growth and development (use of potential)				
12. Trust (in self and others)				
13. Pleasure (enjoyable, fun-filled life)				
14. Power (authority, influence over others)				
15. Responsibility (accountable for important results)				
16. Self-respect (self-esteem, pride)				
17. Social service (helping others, improving society)				
18. Social recognition (status, respect, admiration)				
19. Winning (in competition with others)				
20. Wisdom (mature understanding of life)				
21. Intellect stimulation (thought provoking)				
22. Health (for self, others, and environment)				

TABLE 1-2: Personal Values Worksheet Results

Most Important Personal Values	Importance Weight
1.	
2.	
3.	
4.	
5.	

2 = somewhat concerned
3 = quite concerned
4 = considerably concerned
5 = extremely concerned

Column C

Identify how you would feel if your current satisfaction of this value were significantly *increased.* Use the following scale:

0 = indifferent
1 = slightly happier
2 = somewhat happier
3 = much happier
4 = considerably happier
5 = extremely happier

Column D

Add columns B and C together and place a total in column D. This reflects the *relative importance* of each personal value to you. Place an asterisk (*) next to the five or six personal values that have the highest score in column D.

Review the five or six values you placed an asterisk next to, and ask yourself the following questions:

- As I think back on my experiences (job, career, life), do these values seem to be the most important values for me? If not, what changes do I need to make?
- How do my five or six most important personal values from the worksheet compare with their satisfaction scores in column A? On which values do I want to increase my satisfaction?

Your next task is an extremely important one. From the results of your Personal Values Worksheet and your self-questioning, list below the five personal values that are most important to you. The order of listing makes no difference.

Now you must decide which one of the five personal values is the *most* important to you and place a 10 opposite that value in the Importance Weight column. Compare the other four values to your first choice and assign each an appropriate number from 1 through 9. Ask yourself such questions as, "Is this value about 80 percent as important as my first value?" If so, assign that value an Importance Weight of 8. Continue until all five values have been weighted. Each of your five values should have a different weighting.

You are almost at the finish line. With an idea of what is important to you—deep down important—formulate *a personal goal statement* for your life. Write two or three sentences, starting with, "I want . . . I wish . . . I would like . . ." Putting the pen to paper is important here; the writing makes it real. A rough draft is fine—just get your thoughts on paper.

Congratulations! You have just written a statement that provides the *big* picture, a compass on which to base all decision making—value-based goals that reflect your true wants and needs. Ask yourself when you consider options,

"Will these actions get me closer to my goals, or farther away?" These goal statements will come to define your quality of life.

Quality of Life

Here is an old personal goal statement of mine. Nothing fancy, it's a rough draft of thoughts at the time, but it is written down.

> I want to have a healthy life with family and friends; to earn a living that I have passion for and that doesn't compromise my values of a healthy planet, environment, community, and family. I want to be able to afford to travel, not work extremely long hours, have money enough not to worry about it, feel more spirituality, play more music, spend more time with family and friends.

Look! No farming in there! What I wanted at that point was more economic security and time for family and friends.

A clearly defined goal should define your quality of life. We all have a quality of life: It may be good or not so good, defined or undefined. But we all have some sort of quality of life. The choice is stark between seeking a certain quality of life and just letting it happen.

Whether our quality of life is defined or not, we know when actions are working toward or against it. People will put up with tremendous hardship if they are working toward their quality of life and deeply held goals. Conversely, when things oppose your goals, tolerance is very short.

Here is a simplified example. Suppose my overriding goal was to go from New York City to the top of Mount Washington in New Hampshire. I start out driving from New York, but my car breaks down in Connecticut. I leave the car and start hitchhiking. A brand-new limousine stops and picks me up, but the driver says that he is headed for the Super Bowl and has an extra ticket: Would I like to go? Focused on my goal, I say thanks but decline, get out of the car, and start walking in a beeline toward Mount Washington. Hilly terrain, inclement weather, wild animals, and quirky tourists don't deter me. I keep walking, with my eye on the summit.

Farmers tend to survive very tough economic times because they believe in what they are doing; that is, working toward their goals. Having a defined goal statement makes all the difference in your attitude—it keeps the big picture in sharp focus. At some point after I started farming, I lost sight of my goals, and life was taking a pessimistic turn. At the end of a long season, I had just finished washing my ten zillionth carrot, feeling frustrated from the long hours with low pay, and I was ready to quit farming. *This is it,* I thought. *I hope I lose money this year so I have a good reason to stop.* But a couple of weeks later, after pondering my quality of life and writing down my goals, I found that going to the barn looked like a totally different, and much more satisfying, occupation.

Before I had a clear goal, I was washing carrots simply to wash carrots. Meaning and purpose were absent. Now I wash carrots knowing that I am supporting my family, enabling leisure time, planning for retirement and my kids' future, building community, and living the life that I want to live. Carrot washing was put in *perspective.* To the outside observer watching me in the washroom, nothing had changed—just my mind-set. At some point in the frustrations of my life and farm, I had temporarily forgotten my original purpose in growing carrots. With the big picture now back in focus, those same carrots became my summit of Mount Washington.

Life on the farm has its ups and downs. Things don't always happen according to plan. Frustrations arise. Put your goal statement along some well-traveled route in your life, like on the refrigerator door, and always remember to keep the big picture in focus.

– 2 –

Farm for Profit, Not Production

Chapter 1 explored what I call the "Inner Game of Farming"—in other words, the *no-numbers* approach to business: solar dollars, paradigms, quality of life, and goal setting. If you haven't yet completed the goal-setting exercise, set aside some time soon to do so.

Enough with the fuzzy concepts, though. It's time now to get down to business. How do you actually make money from farming and put real greenbacks into your bank account?

Unfortunately, financial management and record keeping can sound like some of the most boring subjects on earth. Just the term *number crunching* seems to numb the mind, and yet it is one of the most important tools in the farmer's toolbox. For years, I didn't pay enough attention to it, but somehow I survived, owing in part to putting in some very long days over the course of ten years. I knew that I wanted to work less and make more money—a goal that most people share—but in my case, I wanted to work a *lot* less and make a *lot* more money.

A person involved in any other line of business would think it ludicrous that many farmers don't keep track of where the money comes from and where it goes. Every year, farmers may handle large sums of money—$50,000, $100,000, $200,000 or more—yet only have $20,000 net income in a good year, and break even *or even lose money* in a bad year. And that's with working your tail off! Why? Do you think the auto parts store or shoe store runs a business without knowing the numbers?

Figure 2-1 is an actual Schedule F tax form that was filed by a farmer colleague.

The farm's gross sales are quite substantial, more than $230,000, yet the bottom line reveals a very meager net profit—only $366

SCHEDULE F (Form 1040)
Department of the Treasury Internal Revenue Service (4)

Farm Income and Expenses

▶ Attach to Form 1040, Form 1041, or Form 1065.
▶ See Instructions for Schedule F (Form 1040).

OMB No. 1545-0074
1989
Attachment Sequence No. 14

Name of proprietor: COPY | Social security number (SSN)

A Principal product. (Describe in one or two words your principal crop or activity for the current tax year.) | B Agricultural activity code (from Part IV) ▶

C Accounting method: ☐ Cash ☐ Accrual | D Employer ID number (Not SSN)

E Did you make an election in a prior year to include Commodity Credit Corporation loan proceeds as income in that year? ☐ Yes ☐ No
F Did you "materially participate" in the operation of this business during 1989? (If "No," see Instructions for limitations on losses.) ☐ Yes ☐ No
G Do you elect, or did you previously elect, to currently deduct certain preproductive period expenses? (See Instructions.) ☐ Does not apply ☐ Yes ☐ No
If you choose to revoke a prior election for animals, see the Instructions.

Part I Farm Income—Cash Method—Complete Parts I and II (Accrual method taxpayers complete Parts II and III, and line 11 of Part I.)
Do not include sales of livestock held for draft, breeding, sport, or dairy purposes; report these sales on Form 4797.

Line	Description	Amount
1	Sales of livestock and other items you bought for resale	
2	Cost or other basis of livestock and other items you bought for resale	
3	Subtract line 2 from line 1	
4	Sales of livestock, produce, grains, and other products you raised	232,225
5a	Total cooperative distributions (Form(s) 1099-PATR) — 5b Taxable amount	
6a	Agricultural program payments (see Instructions) — 6b Taxable amount	
7	Commodity Credit Corporation (CCC) loans:	
7a	CCC loans reported under election (see Instructions)	
7b	CCC loans forfeited or repaid with certificates — 7c Taxable amount	
8	Crop insurance proceeds and certain disaster payments (see Instructions):	
8a	Amount received in 1989 — 8b Taxable amount	
8c	If election to defer to 1990 is attached, check here ▶ ☐ — 8d Amount deferred from 1988	
9	Custom hire (machine work) income	6,000
10	Other income, including Federal and state gasoline or fuel tax credit or refund (see Instructions)	
11	Add amounts in the right column for lines 3 through 10. If accrual method taxpayer, enter the amount from page 2, line 51. This is your **gross income** ▶	238,228

Part II Farm Expenses—Cash and Accrual Method (Do not include personal or living expenses such as taxes, insurance, repairs, etc., on your home.)

Line	Description	Amount	Line	Description	Amount
12	Breeding fees		24	Labor hired (less jobs credit)	94,295 —
13	Chemicals		25	Pension and profit-sharing plans	
14	Conservation expenses (you must attach **Form 8645**)		26	Rent or lease:	
15	Custom hire (machine work)		26a	Machinery and equipment	5322
16	Depreciation and section 179 deduction not claimed elsewhere (from **Form 4562**)	11,336 —	26b	Other (land, animals, etc.)	5190
17	Employee benefit programs other than on line 25		27	Repairs and maintenance	9943 —
18	Feed purchased		28	Seeds and plants purchased	4734 —
19	Fertilizers and lime	6222 —	29	Storage and warehousing	
20	Freight and trucking	574 —	30	Supplies purchased	15,057 —
21	Gasoline, fuel, and oil	8305 —	31	Taxes	2356 —
22	Insurance (other than health)	7267	32	Utilities (see Instructions)	4993 —
23	Interest:		33	Veterinary fees and medicine	250
23a	Mortgage (paid to banks, etc.)		34	Other expenses (specify):	
23b	Other	12410	34a	Dues, mag	1251
			34b	Boxes Packaging	25,667
			34c	coop fee	22,690
			34d		
			34e		

Line	Description	Amount
35	Add amounts on lines 12 through 34e. These are your **total expenses** ▶	237,862
36	**Net farm profit or (loss).** Subtract line 35 from line 11. If a profit, enter on Form 1040, line 19, and on Schedule SE, line 1. If a loss, you MUST go on to line 37. (Fiduciaries and partnerships, see Instructions.)	366

37 If you have a loss, you MUST check the box that describes your investment in this activity (see Instructions). 37a ☐ All investment is at risk. 37b ☐ Some investment is not at risk.
If you checked 37a, enter the loss on Form 1040, line 19, and Schedule SE, line 1.
If you checked 37b, you MUST attach **Form 6198.**

For Paperwork Reduction Act Notice, see Form 1040 Instructions. Schedule F (Form 1040) 1989

Figure 2-1: IRS Schedule F.

for a full year's worth of work. That's about a dollar a day net profit! How can this be?

An analogy helps illustrate this situation. A traveler was walking through the woods and came across a logger in the process of sawing down a tree. The traveler paused and watched as the logger worked hard at sawing and sawing and sawing, but he was making very little progress in cutting the tree. The traveler approached the logger and asked, "Why don't you take five minutes and sharpen your saw?" To which the logger replied, "I can't! I'm too busy cutting down the tree!"

I have personally wasted lots of time and money, being too stubborn to face up to the reality of the situation. Over the years, I did get a better grasp on which crops were paying for themselves and which ones weren't, but it wasn't until extreme frustration set in that I really sat down and started to figure things out systematically.

What amazed me was my own internal resistance to actually crunching the numbers for this finely tuned financial analysis. It's not that I dislike math; on the contrary, I enjoy it. *I was scared of change,* the change that these numbers might show me. Scared of finding out that farming was indeed unprofitable, that my livelihood and identity might have to drastically change, and that a decade of blood, sweat, and tears would go down the drain.

Like it or not, though, farmers are in the *business* of farming. Aside from the growing, personnel management, and marketing, farmers are also businesspeople. It is the farmer's job to make sure the farm business survives. If you want your farm to make you money, it needs to be profitable. Basic rule of business: Stop doing things that lose money.

Profit = Income – Expenses

Farm for profit, not production. A simple phrase, yet it took me ten years to figure it out. Fuzzy business concepts are brought into sharp focus with the irrefutable business equation: *Profit = Income – Expenses.* Like Einstein's famous equation $E = mc^2$, the simplicity of Profit = Income – Expenses belies its importance. All too often, farmers calculate profit (or loss) only once a year at tax time.

The annual calculation of Profit = Income – Expenses, however, is usually *only an average* of all the various enterprises that make up the farm. Diversified organic farms may have dozens of different parts that make up the whole, whether they are lettuce, tomatoes, and zucchini or grain, hay, heifers, and milkers. Each enterprise has its own unique costs and benefits. The odds of having all the farm's enterprises equally profitable are extremely low. In reality, a farm's different enterprises probably look more like this:

Profit = Income – Expenses

Profit = Income – Expenses

– Profit = Income – Expenses

Profit = Income – Expenses

and so on, *resulting in an average Profit = Income – Expenses for the whole farm operation.* To become more profitable, the farmer must sharpen his or her saw—or pencil, to be exact—and determine which enterprises are more profitable and which are less so, or downright unprofitable. Place more emphasis on the more profitable ventures and less on the less profitable ones. Evaluate the enterprises to see if the profitability equation can be altered by raising the sale price, reducing expenses, or both. Focusing your efforts on the most profitable enterprises and reducing or eliminating the unprofitable ones can only increase your farm's overall bottom line.

Planning for Profit

Let's address the profit side of the equation first. *Farm for profit, not production.* This runs contrary to the common belief that farm profit is defined as what is left over (if there is any left over) at the end of the year. This is a passive approach to profit, letting your income and expenses become the determining factors. Instead, be proactive in your approach to profit. How much do you want to make per year: $20,000? $50,000? $80,000? Set your financial goals for net return and then plan for them.

The dollar amount you want to make from farming is a personal choice, one that adequately compensates for the time and money you invest. It is human nature to base your expectations of what a fair salary is by looking around at what other people earn: brothers, sisters, parents, neighbors, schoolteachers, tradespeople, or business owners, for example. But more important is what you would *like* to make. Write this number down, and remember that it may take you more than twelve months to get there.

For purposes of illustration, let's say that you want to make $30,000 per year net profit. In order to net $30,000 per year, gross sales would have to be at minimum $30,000, and, with typical farm expenses, gross sales might need to be two or three times the desired net profit amount.

Depending on what your current gross sales and net profit are, or if you are just starting out, map out some possible scenarios over a five-year time frame. Let's say that you now net $2,000 from $10,000 gross sales. Remember, Rome was not built in a day.

Start by putting $30,000 net profit as a goal for five years from now (table 2-1). Next, what gross sales will you need in order to net that $30,000 in year 5? It is a tough question; guesstimate for now (this will be addressed later on). Let's use $60,000 gross sales (table 2-2). Continue filling in years 1 through 4, with gradual increases that lead up to your final destination of $30,000/year net profit (table 2-3).

Now you have a rough blueprint of how to achieve your goal of $30,000 net profit. As your farm becomes a more efficient business through better financial analysis, more profit will be squeezed from gross sales, so the ratio of gross sales to net profit may change over time.

So, if we assume that a certain volume of gross sales is needed to generate a particular

TABLE 2-1

	This Year	Year 1	Year 2	Year 3	Year 4	Year 5
Gross sales	$10,000					
Net profit	$2,000					$30,000

TABLE 2-2

	This Year	Year 1	Year 2	Year 3	Year 4	Year 5
Gross sales	$10,000					$60,000
Net profit	$2,000					$30,000

TABLE 2-3

	This Year	Year 1	Year 2	Year 3	Year 4	Year 5
Gross sales	$10,000	$20,000	$30,000	$40,000	$50,000	$60,000
Net profit	$2,000	$6,000	$12,000	$18,000	$24,000	$30,000

net profit, *where will those gross sales come from?* What do you expect to sell, to whom, how much of each product, and at what price? How and where will you grow each crop? These are big questions, but they're easily tackled if broken into smaller pieces.

I like to look at the planning process as a road map of how you get to where you want to go. Once you have planned your destination as a certain net profit, break up the journey into a few smaller segments. I use four stops along the way: the Marketing Chart, the Production Plan, the Map, and the Seedling Calendar.

1. Marketing Chart

Start by breaking down your current or projected sales into a spreadsheet format: accounts you sell to across the top, and the products you sell along the left side, with dollar and production unit (lbs) amounts in each cell. I call this a Marketing Chart. A simplified version for a vegetable farm is depicted in table 2-4. This one-page snapshot of your marketing efforts is tremendously informative.

Totals are tabulated for each crop, and for each account, with a grand total for all sales in the lower right-hand corner. This one chart represents everything you sell, or plan to sell, from the farm. If you are projecting future sales, you can base estimates on past sales or research speculated crops and accounts. Ask produce buyers if they are willing to buy a certain crop from you, and if so, what quantity and at what price. It is not a guarantee that the buyers will indeed purchase from you, but it is a start of a working business relationship. Farmers' market and CSA (community-supported agriculture) sales depend on numerous factors. Again, do some research to get a workable sales figure (find tips for managing a CSA in chapter 5).

TABLE 2-4: Simplified Marketing Chart

	Farmers' Market	Food Co-op	Restaurant	CSA	Other	Total
Beets	8 wks x 25 lb = 200 lb = $400	12 wks x 50 lb = 600 lb = $600	0	4 wks x 50 lb = 200 lb = $400	8 wks x 25 lb = 200 lb = $200	1,200 lb $1,600
Carrots	8 wks x 50 lb = 400 lb = $800	8 wks x 200 lb = 1,600 lb = $2,000	8 wks x 50 lb = 400 lb = $500	12 wks x 50 lb = 600 lb = $1,200	12 wks x 75 lb = 900 lb = $1,125	3,900 lb $5,625
Lettuce	10 wks x 60 hds = 600 hds = $1,200	10 wks x 120 hds = 1,200 hds = $1,200	10 wks x 120 hds = 1,200 hds = $1,200	16 wks x 50 hds = 800 hds = $1,600	10 wks x 120 hds = 1,200 hds = $1,200	5,000 hds $6,400
Potatoes	6 wks x 100 lb = 600 lb = $900	10 wks x 200 lb = 2,000 lb = $3,000	0	6 wks x 100 lb = 600 lb = $900	6 wks x 100 lb = 600 lb = $900	3,800 lb $5,700
Total	$3,300	$6,800	$1,700	$4,100	$3,425	$19,325

Crop prices for retail markets are often higher than those for wholesale accounts, as reflected in the above chart. The list of items you sell may be quite long—possibly forty or more. But once you've created your first Marketing Chart, subsequent ones are a snap.

2. Production Plan

After thoughtful contemplation about who will buy what product at what price, totals are determined for each product to be grown, as shown in the Marketing Chart's right-hand column. These totals are used to create the farm's Production Plan, a detailed account of how to produce what you hope to sell (table 2-5).

Yields for each crop can be determined from past records, other growers, a Johnny's Selected Seeds or other wholesale grower's catalog, or *Knott's Handbook for Vegetable Growers.* Beds may be any size; I chose the 350-foot length (and 6 feet from aisle to aisle, which works with common tractor wheel spacing) as a workable size. There are twenty beds per acre, making number crunching much easier.

3. Mapping It Out

Now that you know how many beds or acres you'll need to grow for your projected gross sales, it is time to see how this projection works with your available tillable land base. Add up all the beds from the third column of the Production Plan. Do you have enough land ready for your projected production? Where will each crop go, and when? Draw an outline of all your fields and make a farm Map (table 2-6).

Draw your map on a big piece of lecture

TABLE 2-5: Production Plan

Total beds = 16

Crop	Yield/350' Bed	# Beds Needed	Proj. Gross Sales	Seed Needed	Planting Dates	Notes
Beets 1,200 lb	600 lb (conservative)	2 beds (1 early + 1 late)	$1,600	16,000 RedAce 16,000 Gold	5/1, 6/1	15 seeds/foot
Carrots 3,900 lb	1,000 lb	4 beds (1 early + 3 late)	$5,625	27,000 Nelson 27,000 Bolero	5/1, 6/1	25 seeds/foot w/ scatter shoe
Lettuce 5,000 hds	830 heads (marketable) 3 rows x 12" 80% pick	6 beds 8 plantings 0.75 bed/ planting	$6,400	2,400 Two Star 1,200 Vulcan 1,200 Cocarde 1,200 Ermosa	5/1, 5/15, 6/1, 6/15, 7/1, 7/15, 8/1, 8/15	750 plants needed per planting to sell 625 hds
Potatoes 3,800 lb	950 lb	4 beds	$5,700	100 lb gold 100 lb red	5/1	Red:gold 1:1 Single row/bed 1:20 seed:crop

paper and fill in as much information as you want. Hang your map on the wall in a place where it will be seen frequently. It is a great working tool. Be very specific. When planning where to put each crop, take into consideration crop family rotation, access to water, frost sensitivity (crops grouped for row coverings), deer pressure, early versus late planting and harvests, what the field will look like when crops are harvested and tilled under, and cover-cropping schemes. Some crops can be double-cropped on the same land: early spinach followed by late lettuce, or early peas followed by late beets. Update your map as the season progresses. All my maps for the past five years are layered on my office wall for quick reference for this year or previous plantings. The map is a visual farm plan of everything you grow, a real wealth of information. Once it's complete, I refer to my map at least once a week and follow the plan.

TABLE 2-6: Map

← 350' long beds →
1. **Beets:** 1 bed 5/1, Red Ace and Gold, 3 rows/bed 2. 1 bed 6/1, Red Ace and Gold, 3 rows/bed
3. **Carrots:** 1 bed 5/1, Nelson, 3 rows/bed 4. 1 bed 6/1, Nelson, 3 rows/bed 5. 1 bed 6/1, Bolero, 3 rows/bed 6. 1 bed 6/1, Bolero, 3 rows/bed
7.–12. **Lettuce:** 0.75 bed every 2 weeks: 5/1, 5/15, 6/1, 6/15, 7/1, 7/15, 8/1, 8/15 3 rows/bed at 12" spacing 6 beds total 2 green:1 red:1 red oak:1 Boston each planting
13. **Potatoes:** 1 bed 5/1 Red Norland 14. 1 bed 5/1 Red Norland 15. 1 bed 5/1 Carola 16. 1 bed 5/1 Carola
17.–26. **Cover Crop:** 10 beds (0.5 acre) 5/15 buckwheat 9/15 fall oats Farm road

TABLE 2-7: Seedling Calendar

Crop	3/1	3/8	3/15	3/22	3/29	4/5	4/12	4/19	4/26	5/3	5/10
Lettuce					750		750		750		750
Onions	5,000										
Broccoli				700		700		700			

4. Seedling Calendar

The last piece necessary to carry out your production plan is a Seedling Calendar. On May 1, you need to plant out 750 heads of lettuce. Where will those plants come from? The local convenience store? I don't think so. Unless there is a readily available local supply of the seedlings you intend to plant, plan on growing them yourself. Unlike direct-seeded crops, transplants have a much shorter time frame in which you can get them in the ground. In order to have 750 lettuce plants ready on May 1, you'll need to sow lettuce seeds in planting trays four to five weeks earlier. The same goes for all the subsequent plantings of lettuce, as well as any crops that are grown from transplants.

Look at your Map and note any crops that are grown from transplants, then figure when you need to start seeds for them. Place the number of transplants needed on the Seedling Calendar (table 2-7).

Summing It Up

To review the overall planning-for-profit process, first determine what you hope to sell (and to whom) for the upcoming season and create a Marketing Chart. Next, take the total amounts of each crop you will need to produce from the Marketing Chart and set up a Production Plan. The Production Plan describes the details of growing the crop and determines the amount of land you'll need for each crop that you plan to sell. These amounts (in beds or acres) are portrayed on the Map as a visual picture of your fields. And finally, a Seedling Calendar schedules the growing of any crop grown from transplants.

Always start by determining what you think you can sell, and then plan on producing that amount. Farming can be a low-margin business, so efficiency is paramount. Growing a crop that is unwanted and selling it at fire-sale prices—or worse yet, tilling it in just before potential harvest—is an unwise business practice. Almost all costs of production have been paid, with little or no income to offset them.

Thoughtful and systematic planning will result in optimal efficiency and farm profits. Planning for profit methodically shows the route to obtaining the gross sales that will in turn generate your desired *net* profit. The next step is to figure out how to get the best net return from gross sales.

– 3 –

Discovering Profit Centers

Let's revisit the old business equation Profit = Income – Expenses. The previous chapter showed how overall farm profitability comprises many different crops or enterprises, each with its own profitability equation. All these individual farm enterprises generate profits (or losses) and are combined to create an average annual profit (or loss) for the entire farm. Unless we know how each part contributes to the whole (carrots versus beets versus blueberries versus eggs versus sheep . . .), we are operating in the dark and have a hard time figuring out how to increase overall net profit. The chances are slim that all farm endeavors contribute equally to the farm's overall profit. Some crops or enterprises are more profitable than others, and some may even be losing money. Once we're armed with the knowledge of which farm sectors make money and which do not, increasing overall farm profitability is a snap.

Chapter 2 also tackled the concept of planning for profit, determining first how much you want to make in net profit for the entire farm, and then planning production to get there. This chapter delves deeper into the Profit = Income – Expenses equation to help you reach your desired profit goals.

So, if I know what I want my profit to be, there are only two other variables to deal with: income and expenses. How hard can that be? Let's start with income.

Tracking Income

Farm sales can come from many different sources and are tracked in different ways. The Marketing Chart from the preceding chapter portrays retail sales from a farmers' market, wholesale sales to a food co-op and a restaurant, and CSA sales. A farm stand is another common sales route. To truly quantify the income variable of the profit equation for each crop or enterprise, we need to know how much of the product leaves the farm and what its sales value is.

Sales to wholesale accounts (stores, restaurants, growers' co-ops) are often the most easily monitored for the simple reason that both you and the buyer are accustomed to using invoices as a standard business procedure. Duplicate invoices are filled out listing each product and price, and they're signed by the buyer upon delivery. The buyer keeps one copy, and you keep the other. Regardless of whether you get paid upon delivery or not, your copy of the invoice is your record of the items, in quantity and dollar amount, that left your farm. Use of duplicate invoices creates an easy-to-follow paper trail. Total sales for each product or account can be either tabulated by hand or input into a computer.

But unlike wholesaling, other sales venues are not accompanied by such a common practice as invoicing. Often at the close of a farmers' market, growers are more concerned with packing up and getting home than knowing exactly how many cabbages were sold. Calculating total dollar sales may be a priority, but individual crop sales remain unmonitored.

Tracking each crop's sales at every farmers' market need not be either difficult or time consuming; all that it requires is getting yourself into the habit of gathering some data. Two pieces of information are needed: a beginning inventory of what items (and their amounts) were brought to market, and an ending inventory of what is left at the end of market. Five minutes before packing up at market, I take my list of items that I brought and record what is still remaining on the tables. I estimate weights of root crops in bushel baskets, or tomatoes in boxes. Counting cabbages and bunched crops is easier and more precise. All I do is record what's left over, and then pack up to go home. Time-wise, recording your beginning inventory on a sheet of paper on a clipboard may take you ninety seconds to a couple of minutes at each market. Marking off ending inventory takes another one to three minutes. There's no need to do the subtraction math now—in fact, I often do that in the off-season when I have more time. When I count my money apron, I write the total sales figure on the inventory sheet for future reconciliation. Then I set aside my farmers' market sales inventories until I have time to tabulate them.

Tracking farm stand sales is similar to the farmers' market procedure. Record what is at the stand at the beginning of each day (or week), add in any product brought in during the day (or week), and finish by noting ending inventory. Modern cash registers can simplify data collection if you assign individual keys to different products, like tomatoes or cabbages, but paper inventories are nonetheless easy and effective. Again, the math of tabulating each crop's sales can wait until later if need be. None of this data collection is that time consuming, but it does have to occur on a regular basis. It's like brushing your teeth. Your dentist would much prefer you to brush a little bit each day rather than waiting until just before your six-month checkup and brushing for three hours nonstop.

CSA income is another easy-to-track sale, and all you need is paper and pen. Every week's CSA share is made up of what's available that week and what the customer likely would prefer. Cost of the share is another issue, and we will be looking at that in chapter 5. CSA growers determine which crops, and which amount of each, will constitute a share. For instance: "August 15, 2008, 1 lb beets, 2 lb tomatoes, 1 lettuce, 2 cukes, 1 eggplant, PYO herbs, 20 stems flowers . . ." Simply record this CSA share information in a loose-leaf or spiral-bound notebook, to be compiled later on with other sales accounts.

All income coming into the farm should be accounted for. Data from all these sales outlets are now compiled into a Sales Spreadsheet (table 3-1), in the same format as the Marketing Chart. Accounts are named across the top, with different crops or enterprises listed along the left side.

In a perfect world, figures on the Sales Spreadsheet would exactly match your projections from the Marketing Chart. But, alas, farming is not an exact science. Plan as best you can, and then follow the plan.

The Sales Spreadsheet is a valuable one-page summary of all your sales income. You can monitor trends in each account, and in each crop. Plus, you have a grand total for all sales. Comparing sales from year to year, or even more often, is enlightening and helpful for future planning. For ease of understanding, I omitted putting in actual quantities of each crop (240 pounds of beets, for example), but that information is very useful and I recommend you track it. Computers make this task a snap in the long run, but even with pencil and paper only a few hours per year will be required.

That's it for the income part of Profit = Income – Expenses. All that remains is expenses.

TABLE 3-1: Simplified Sales Spreadsheet

	Farmers' Market	Food Co-op	Restaurant	CSA	Other	Total
Beets	$480	$650	0	$400	$150	$1,680
Carrots	$890	$2,100	$470	$1,000	$600	$5,060
Lettuce	$1,310	$1,140	$1,280	$1,400	$960	$6,090
Potatoes	$1,100	$3,250	0	$980	$850	$6,180
Total	$3,780	$7,140	$1,750	$3,780	$2,560	$19,010 Grand Total

Tracking Expenses

You'd think keeping track of expenses would be simple: Checks are written, receipts collected, and actual pieces of paper are there to prove it. True enough. I buy seeds, get a bill for them, and write a check for payment. When I make purchases with cash, I still have a bill for documentation. (See chapter 7 for possible foibles of using cash.) My paid bills are easy to break down into various expenses types, to then plug into the Profit = Income – Expenses mantra for each separate crop. I know how much my beet seed cost, how much I paid for rock phosphate, and the expense for each 25-pound plastic bag.

But what about all the expenses in between purchasing the seed and packing the beets into their 25-pound traveling bag? Planting the seed, weeding the crop, irrigating when needed, and harvesting constitute the majority of costs in raising a crop and must be accounted for. In a sense, some of those expenses *are* documented—in the form of a weekly paycheck to employees. But the paycheck is just a total of all hours worked on numerous tasks on numerous enterprises. Paychecks do not differentiate time spent working on beets, carrots, potatoes, and blueberries. Furthermore, your labor in the field as owner-manager isn't even recorded in a pay stub—so how do you figure out where *your* time was spent?

Systems need to be in place for tracking those expenses that skirt the conventional itemized paper trail. Whereas paid bills are easy to break down into expenses for different enterprise budgets, your own labor and your employees' time require extra documentation. At season's end, when I have time to see what was actually making money and what wasn't, I want to have accurate information at hand from the previous growing season to create individual crop budgets. The labor input for each crop spans months (or years) of work, often sporadically, in bits and pieces. Recording the information isn't all that time consuming, but it has to be done on a regular basis throughout the year.

The Indispensable Crop Journal

One of the most important books on my farm is the Crop Journal. This is just a basic pocket folder with loose-leaf pages inside, one page for each crop or enterprise, arranged alphabetically. Anytime a task is performed on a crop, it is recorded on the appropriate page in the Crop Journal. It's simple and quick to do once you get in the habit. Years ago, my crew chief was startled when I mentioned how few farmers do this. She said, "It's so easy!"

All information necessary for later budgeting (and future planning) is written down in the Crop Journal. Preparing soil, seeding, cultivating, and harvest are all recorded. A sample page at the end of the season is shown in table 3-2.

Information is recorded *contemporaneously*, or, in plain language, as it happens. I usually do it at the end of each day. It doesn't take long, especially once you get into the habit. I have certain employees carry a pen, some

TABLE 3-2: Sample Crop Journal: Carrots				
¼ acre (five 350' beds), 3 rows per bed				Location: Lower field
Date	**Task**	**Labor**	**Tractor/equipment**	
4/23	Spread 1 load compost (4 yards or 4,800 lb)	.5	.5	(at 2 hours/acre)
	Spread 75 lb SoPoMg	.25	.25	(at 1 hour/acre)
	Spread 200 lb bagged poultry compost	.25	.25	(at 1 hour/acre)
	Disked 1x	.25	.25	(at 1 hour/acre)
4/24	Chisel 1x, bedform 1x	.5	.5	(at 1 hour/acre each)
4/24	Seed 25,000 Napoli, planter hole #10			
	Seed 100,000 Bolero, hole #8	1		
5/7	Flame weed	.5	.5	(at 2 hours/acre)
5/25	Cultivate with baskets	.25	.25	(at 1 hour/acre)
6/6	Cultivate with baskets	.25	.25	
6/10	Irrigate: set up, run, and put away	1.5	2	
6/15	Hand-weed crew	16		
6/16	Finish hand-weed	4		
6/18	Cultivate with baskets	.25	.25	
7/6	Cultivate with sweeps	.25	.25	
7/9	Irrigate	1.5	2	
7/16	Cultivate with sweeps	.25	.25	
7/18	Hand-weed crew	10		
8/1	Cultivate wheel tracks	.25	.25	(at 1 hour/acre)
8/15	Last hand-weed	2		
9/15	Bedlift 2 beds	.75	.75	(start to finish)
9/16	Harvest: 55 bushels	18		
10/2	Bedlift 3 beds	1	1	
10/2	Harvest: 86 bushels	29		
10/4	Disk 1x	.25	.25	
10/6	Seed oats with Brillion	.75	.75	
10/7	Wash and pack	29		(Rate: seven 25-lb bags/hour)
Note: Because more than ¼ acre of land is prepared at once for efficiency, I use a per-acre rate. If you're preparing only ¼ acre, labor and tractor times will be higher.				

paper, and a watch as part of their job. I find that my employees like the added responsibility of tracking crop production details and the focus on the economic side of farming. When keeping track of time, employees are reminded that farming is production work, and that the farm earns money by the piece, not by the hour. Managerially, I tend to organize work in blocks—everyone weeding carrots in the morning, trellising tomatoes in the afternoon. Notice that the information in the Crop Journal is only recorded, and not yet tabulated. The math of each crop's budget comes later, but the information needed for it is documented as it occurs. The once elusive component of labor now can be easily calculated for individual crop budgets. I often don't bother with listing prices for material inputs such as seed or fertilizers in the journal because I already have my paid bills as a record for those expenses.

I pay particular attention to *rates:* How long does it take to seed one bed? How many bushels are harvested per hour, or per bed? Rates involve at least two different parameters, such as quantity/area, quantity/time, and time/area. These rates can be extrapolated to larger amounts, and they are very useful for planning and crop budgeting. Many of the hour amounts that I listed on the carrot page are from rates for a larger area. That's because when I plan to cultivate my ¼ acre of carrots by tractor, I plan to use the tractor for other crops as well. It is more efficient to bundle tasks than to repeat steps for each crop given the setup, travel, and takedown time with each tractor implement.

I also keep a page labeled STANDARDS in my Crop Journal for rates of farm tasks that are done repeatedly. Updated often, this information is a perfect way to give employees an idea of what is expected, and to help you as a manager plan what to tackle on any given day. Some examples of production tasks are: bunching cilantro (150–200 bunches/hour), picking spinach (3–4 cases/hour), and filling 3-inch pots (60 trays/hour). The standards page is very useful information, and hard to find elsewhere.

An important thing about record keeping in the Crop Journal is that it shows information for each crop in hindsight. The numbers from this year will yield pertinent information at the end of the year, and for years to come. (The data can also be used to project future budgets of possible new enterprises.) The chronological log of each input for a given crop can now be used to accurately track expenses in the Profit = Income – Expenses equation.

Creating a Simple Crop Budget

Accurate tracking of income uses simple systems: invoices, farmers' market inventory sheets, farm stand inventory sheets, and a CSA log of each share. Expenses are monitored by the paid bills/checkbook register and the Crop Journal. The business maxim Profit = Income – Expenses can now be calculated *for each individual crop or enterprise*—not just the overall average for the entire farm. Here is where the rubber meets the road of profitability. Once you have this

Beds of carrots. Credit: Vern Grubinger

information, you can uncover the true costs of production for each crop and see what kind of profit you made (or didn't make) at current prices.

There were a lot of surprises for me when I first did this in-depth analysis. For years, I had gross sales of around $5,000 per acre *average*. Some crops bombed; others did well. I eked out a meager profit year to year but knew I was working too hard for the net return I was getting. Concurrent with the big exhale at the season's end, I yearned to know the true costs of production for each crop. Only by tracking all the variables did I finally get the information I needed.

I began by constructing simple crop budgets that didn't allocate fixed expenses for telephone, advertising, taxes, mortgage interest, insurance, electricity, and people accidentally driving over irrigation pipes. These expenses are spread out over the whole farm operation and can be viewed as constants. Lacking such fixed expenses, of course, my bare-bones crop budgets didn't tell me *exactly* my true costs of production, but they did allow me to compare crops side by side to rate their relative

TABLE 3-3: Simple Crop Budget: Carrots

Five 350' beds, 3 rows/bed
Location: Lower field
Not including fixed costs; for comparing crops only

Expenses	Labor Cost	Machinery Cost	Product Cost
Prepare soil			
Spread 1 load compost	$6.28	$2.50	$100.00
Spread 75 lb SoPoMg	3.14	1.25	18.00
Spread 200 lb bagged poultry compost	3.14	1.25	40.00
Disk 1x	3.14	1.25	0
Chisel and bedform beds	6.28	2.50	0
Seed/transplant			
Plant 125,000 seeds	12.55	0	79.00
Cultivation			
Flame weed	6.28	2.50	8.00
Cultivate with baskets 3x	16.73	3.75	0
Cultivate with sweeps 2x	6.28	2.50	0
Irrigate 1x	18.83	10.00	0
Hand-weed 3x (32 hours)	401.60	0	0
Cultivate wheel tracks 1x	3.14	1.25	0
Harvest			
Bedlift 2 beds	9.43	3.75	0
Harvest (47 hours)	589.85	0	0
Wash, sort, pack (200 25-lb bags, 29 hours)	363.95	0	50.00
Post-harvest			
Disk 1x	3.14	1.25	0
Seed 25 lb oat cover crop	9.43	3.75	22.00
Total expenses	$1,463.19	$ 37.50	$317.00 = $1,817.69

Income			
Sales	**# of 25-lb Bags**	**Price/Bag**	**Total $**
Retail	40	$43.75	$1,750.00
Wholesale	160	25.00	$4,000.00
Total sales			$5,750.00
Net profit per ¼ acre	$5,750.00 – 1,817.69 = **$3,932.31** without fixed costs		

profitability. Such ratings gave me incredibly useful information that I could use to improve my bottom line. I could stop growing the crops on the bottom half of the profitability scale, and instantly my overall farm profit would rise. I could also examine each crop's Profit = Income – Expenses equation and see if I could tweak the profit by either raising prices or decreasing costs—or both. The world of possible profits unfolded before my eyes.

Table 3-3 shows a streamlined version of a crop budget for carrots. I took information primarily from my Crop Journal, with materials costs verified from paid bills. The labor rate is $12.55/hour; the tractor with implement rate, $5.00/hour. (These rates, and other costs, are covered in detail in the next chapter.)

That net profit of $3,932 per ¼ acre really opened my eyes. That would be a net of $15,728 per acre! Whereas before I was averaging *gross sales* of $5,000 per acre, here was a crop netting more than $15,000 per acre. Yikes! I never would have thought those profits were possible until I took the time to figure it out. I started to understand that real farming takes place from the neck up.

Keep in mind that fixed expenses were not accounted for yet, but things sure were looking positive. If I could grow a mix of crops that were this profitable, a mere 2 acres would net me $30,000 before subtracting overhead. I didn't need to grow large amounts of any one crop as long as all the crops I raised were moneymakers. And if demand was huge for any one single profitable crop, I could focus my efforts on growing more of that one crop to saturate the market.

But how did the $15,000-per-acre return for carrots compare with returns for all the other crops I grew? Way above the rest? In the middle of the pack? I needed to know other crops' profitability to see where each stood relative to the others. I took my current top sellers and created a crop budget for each. After I calculated the carrot budget, much information was already at hand for use in subsequent crop budgets. Many growing tasks are similar, reducing the amount of time necessary to construct other budgets. I spent a few hours in front of my Crop Journal, checkbook, seed catalog, and calculator, and voilà! I now had a list of all the top sellers in order of their profitability per acre. It didn't matter how much acreage each crop claimed. Some were only 1/12 acre, some ⅛ acre, but extremely high on the profitability index. Small-area crops are easily marginalized, but not after we measure their stature as net income generators. Piecing together 2 acres of small but very profitable crops nets you more than one barely profitable crop on the same piece of land. And remember, a high sales price doesn't guarantee a high net return. Gross sales and net profit are unrelated! Well, almost unrelated. Remember, *farm for profit, not production.*

Index of Profitability

The single sheet of paper that lists all your crops or enterprises in order of net return is priceless. This

Index of Profitability drives management decisions from planning to harvest. New markets are sought for top earners, and sale prices and costs are scrutinized for bottom-rung profit laggards. Some crops are dropped altogether, and new ones contemplated. Highly profitable crops may become even more profitable if costs can be reduced and/or sale prices edged upward. Once a farm manager is privy to such useful information, bottom-line profits for the entire farm operation rise dramatically. Be profit-driven, not market- or production-driven.

Many farmers feel that they must grow certain crops that are less profitable (loss leaders) to keep customers coming back for all the other, more profitable crops. My advice is to challenge that paradigm. How hard is it to make money overall when you are purposely growing crops that lose money? How important is that loss leader? Will loss of that net return be offset with more sales of highly profitable crops? Can you sell those higher-value crops without the loss leader? Your attention should remain on your net return, not gross sales. For instance, I sell at a farmers' market and have had a CSA for years, yet I grow no sweet corn. Sweet corn was a divot on my road to profitability, so I stopped growing it. When customers ask why I don't grow sweet corn, I tell them that I have a hard time making money growing it given production costs, deer, raccoons, and crows. I recommend that they buy it, as I do, from my neighbors, who feel that corn is worthwhile for them to grow. I find that customers continue to support us and purchase the crops that we do grow.

The last thing I want to do is brag, or sound like some sensational article that says you can make a cagillion dollars per ¼ acre. I just want you to realize that all this seemingly confusing array of record keeping and analysis is worth the effort. If I offered you a job requiring two days' work and paid you $10,000, you'd be hard-pressed not to take it. That's what I paid myself the first year I did it. Anyone with a going business can do the same to some degree. So go sharpen your pencil.

– 4 –

Profit Time: Crop Enterprise Budgets

Crop budgets are vital tools that allow you to analyzc and compare different crops side by side. And this kind of analysis is the central factor in determining which crops are more profitable than others, and which crops *may actually be losing money.* All of these individual parts make up the farm business as a whole and normally only get averaged out in a year-end profit (or loss). Now the pieces of the profit puzzle are crystal clear:

Profit = Income – Expenses
Profit = Income – Expenses
– Profit = Income – Expenses
Profit = Income – Expenses

and so on . . . All of these can now be averaged to yield a single year-end profitability equation:

Profit = Income – Expenses

Once each crop's profitability is uncovered, crops can be rated relative to one another for their contribution to the farm's overall profit. This Index of Profitability is a primary management tool to increase farm profits. Highly profitable crops can be emphasized, and less profitable crops analyzed for possible improved earnings, or dropped altogether.

Chapter 3 covered simple budgets using production costs. Now we need to incorporate fixed expenses, the very real overhead costs that the farm must bear. Insurance, telephone, taxes, land rent, interest, utilities, office supplies, and advertising all add up to a sizable sum. Taking into account *all* farm expenses reveals the *true* costs of production for each crop. Calculating these comprehensive budgets requires some extra math up front, but once accomplished this effort is spread out over numerous crop enterprise budgets, for both present and future benefit. The first budget takes the longest; preparing subsequent budgets is much quicker.

Crop budgets are unique to each farm. All farms bring to the table a certain combination of factors including experience, skills, location, infrastructure, natural resources, markets, and access to services, to name a few. One farm's most profitable crop may not be another farm's highest earner. Some costs of production are easy to determine, like the cost of seed or fertilizer (from invoices), and some are more difficult to evaluate unless records were kept during the growing season. Labor inputs to a crop, like weeding and harvesting, are often only guessed at, yet they are often the largest costs in crop production. See the description of the Crop Journal in chapter 3 for tracking these elusive crop costs.

Prior to calculating the comprehensive (all-encompassing) costs for each crop, certain farm production expenses need to be determined, such as:

- The average hourly employee labor rate.
- The farmer's hourly rate.
- Total overhead costs.
- What each size flat grown in the greenhouse costs.
- The cost per hour of running each tractor and implement.
- The cost of setting up and running an irrigation system per unit area per use.
- Delivery costs.
- Farmers' market fees and other marketing costs.

Worksheets 1 through 4 precede the actual crop budgets at the end of this chapter because information in the worksheets is needed for each crop budget. This is the extra math I referred to earlier. These worksheets show how to calculate many of the production expenses listed above. These costs will then be assigned to each crop enterprise budget. All overhead expenses incurred on the farm need to be accounted for, from advertising to utilities. After these outlays are determined, you can enter them into each actual crop budget. Not all lines in the budget need to be filled in; crop needs vary. Expense categories are listed as a reminder in case you forget an item or two. Brief notes accompany numbers whose origins may not be obvious.

This all may sound overly complicated and overwhelming, but stay with me: These tasks become easier after you've tackled them for the first time.

The total area planted in each crop varies on the farm, so to compare crops side by side for profitability, all budgets are calculated for the size of the actual planting and then extrapolated

Vegetable farm with greenhouse. Credit: Vern Grubinger

to a standard unit area, such as a certain length bed, or by the acre. The following crop budget examples are all based on two 350-foot-long raised beds, which equals 1⁄10 acre.

Assumptions

Farms, like most businesses, are as different and unique as the people who run them. A "typical" organic vegetable farm does not exist. Many variables are involved in each farm business, making a one-size-fits-all model difficult. Still, for comparison purposes we have to start somewhere. Thus I've made some assumptions here for illustration purposes.

1. Operation Size

For simplicity, numbers in this sample workbook are based on a sole proprietorship organic vegetable farm with 5 acres in cultivation and two 96-foot-long plastic-covered greenhouses. One greenhouse produces seedlings; the other produces greenhouse tomatoes grown in the ground. Five acres of cultivated land grows a variety of mixed produce. Operations this size commonly use one or more tractors and an array of implements. If your farm is less mechanized, it is easy to custom-fit the budgets to your operation. Different economies of scale occur at smaller and larger farm operations. Time is involved in setting up and putting away equipment as well as its actual use in the field. In a 5-acre operation, the scale of various tasks (preparing beds, cultivating, irrigating, and so forth) will likely be in the ¼- to 1-acre size. Times designated for labor and machinery in the budgets reflect this. The 1/10-acre size depicted in the crop budgets can be reliably scaled up to an acre (sometimes more) without skewing expenses, but sales prices may decline with greater volume if you are selling to large wholesalers.

2. Standard Bed Length

Lengths of beds depend on the shape and size of any given field. I show all crops in two 350-foot-long raised beds. Raised beds are 42 inches across the top growing area, and 6 feet wide from aisle center to center (compatible with common tractor tire spacing). This length is not unusual for a vegetable farm, and twenty of these beds add up to exactly 1 acre, for easy figuring. If you have longer fields, the 350-foot bed can be doubled to 700 feet, with ten beds per acre. Designate a standard-size bed for your farm that best fits all your fields and typical planting size. To calculate how many of your beds make up an acre, multiply your bed length (in feet) by the overall bed width (from bed center to bed center). This is the amount of square feet in one bed. Next, divide the area of an acre (43,560 square feet) by your bed area and round the result to the nearest whole number. This is the number of beds per acre, and it is very useful in planning and budget work.

3. Labor Rates

All labor is calculated at $12.55/hour ($10/hour plus workers' comp, taxes, and unassigned time like meetings and cleanup; see worksheet 1). When the farmer works on crop production, the hours that he or she contributes are counted as a cost even though no paycheck is written. *This way all production hours are accounted for.* If the farmer breaks a leg and cannot work in the field, his or her hours can be hired out. When working in the field, the farmer is "paid" a base wage similar to the employee wage, but it is the farmer who also keeps any profit for the crop (or suffers any loss). To be clear, if the farmer never worked in the field, the only money made would be the crop budget net profit. If the farmer *does* help in producing the crop, compensation includes any net profit *plus* the hourly wages worked in the field (no paycheck, but hours are recorded in the Crop Journal).

Farms are all too frequently subsidized by undercompensating the farmer's time. Self-exploitation is not a sensible or sustainable business model. We all have only twenty-four hours per day, and some of that is required for resting and nourishment. Of the hours left for work, we need to make the best use of our limited time. Why squander this precious resource working on something that doesn't pay? Treat your time as you would any employee's—with fair compensation. Furthermore, if you do break a leg, get sick, or have to attend a family gathering, the farm remains profitable even while you are paying an employee to replace your time. In the big picture, the hours a farmer works each year in the field producing crops can be substantial. These hours are rewarded financially even if the crop yields a zero net profit—because the farmer's hours are counted as a production cost, even though no paycheck was written.

4. Overhead Appropriation

Overhead costs consist of expenses that are spread out over the entire farm business (such as telephone, office supplies, and insurance) and represent a significant cost to each crop budget. I also enter an overhead cost not common in crop accounting: the labor hours required for business management, office duties, and maintenance. These tasks take time, usually the farmer's, and often go unaccounted. I allot a few hours each week for this work, even though some parts of the year demand more time than others.

The total cost of overhead must be divided and allocated to each crop budget or farm enterprise—otherwise off-farm income would be necessary to pay for them. Overhead costs are divvied up according to the percentage of the individual crop budget's acreage or sales. This is a somewhat subjective allocation. You may opt to assign more overhead to a particular area, such as greenhouses, and less to other areas, like field crops. Do your best to be accurate in spreading out the total overhead costs. The important point is that all overhead costs are assigned *somewhere*. The crop budgets provided in these pages apportion three-quarters of the overhead costs to the 5 acres of field crops and one-quarter for the two greenhouses.

5. Greenhouse Costs

Two different greenhouse operations are portrayed: one with benches for bedding-plant production, and the other for in-ground tomato growing. The bedding-plant greenhouse analysis is more involved (worksheets 2 and 3) and yields great information for hard-to-find greenhouse production costs. The cost of transplants is entered into each appropriate crop spreadsheet. The greenhouse tomato budget is included with the other crop budgets.

6. Tractor and Implement Costs

The costs for tractors, implements, and irrigation on worksheet 4 are based on personal experience. In brief, the formula to figure the hourly cost of using a tractor is:

total cost of tractor, divided by its
years of useful life, plus annual repairs
and annual fuel expenses, divided by the
number of hours used per year.

For instance, a tractor used fifty hours per year often costs a lot more per hour than a tractor used three hundred hours per year.

Tracking various implements' costs is similar to tracking the cost for tractors, but without the fuel expense. Some implements have lots of moving parts (combines, manure spreaders) and cost more than implements like a bedlifter, which has no moving parts.

7. Irrigation Costs

Irrigation cost takes into account the annual equipment and any repair expense (similar to tractors and implements) and also time for setup, running, and taking down the system for the area you water each time. If the system irrigates an acre, the cost can then be prorated to each bed, per use.

The complex and unique nature of a whole farm business makes it difficult to distribute some specific costs in these pages. Delivery costs and farmers' market costs are calculated in worksheet 1 to serve as an example and are subdivided into each different budget. Note how much it costs for two people to load, set up, staff, and travel for *one* farmers' market—a surprising $246. That cost is constant, irrespective of what sales you make. Gross sales of less than the $246 is unprofitable—if you find that to be the case, then wholesaling the crops might be a better option for you. Similarly, delivery expenses are the same whether your truck is full or empty. Pay attention to these costs. Remember, your time is valuable and should be spent in the most profitable way.

Conclusion

As mentioned earlier, weeding and harvesting can account for a large portion of a crop's production costs. This is clearly shown in some of the crop budgets. What is striking, though, is how small a role is played by other tasks, such as bed preparation or seeding. Overhead costs, on the other hand, can be significant and must be accounted for: They are factored in near the end of the budget. Crops that have low gross sales or net profit per bed often become unprofitable when overhead costs are figured into the Profit = Income – Expenses equation. The sales and expense figures for these unprofitable crops ought to be scrutinized for possible improvement in profitability, or else culled from the crop mix.

Most farmers have preconceived notions as to which crops are the most profitable. This bias may originate from other farmers, high demand in the marketplace, or simply an inherent affinity for certain crops. "I just love growing carrots, dang it!" Or "Lettuce is easy to pick, and I get $2.25 at market. And people love it!" While all these quotes may ring true, crunching numbers for each crop is what shines a bright light on *real* profitability.

My favorite crops' rank quickly changed once I unearthed their contribution to my business survival. Vegetables that I loved to eat and were in high demand became less palatable to grow once I realized how little they contributed to my overall bottom line. Conversely, minor crops that I would never have considered a staple in my diet turned out to be cash cows. As you leaf through the crop budgets in this chapter and in the Appendix, you'll notice a wide diversity in net returns. Some of the variation stems from the current sale price, and some from production costs. Of the production expenses, some are similar across all budgets, like soil preparation, seeding, and, to a lesser extent, cultivation. Transplanting, weeding, and harvesting, however, differ more from crop to crop. Questions arise: Is it better to transplant or direct-seed? Should I mechanize transplanting? Are short-term crops better than those with an extended harvest? Is flame weeding worth it?

While there are no short answers to those questions, I'll offer some general thoughts. Since weeding is potentially a big percentage of expenses for some crops, you should direct your focus here. By transplanting crops into a freshly made bed, you get a big jump on the weeds, eliminating many hours of hand weeding, and enhance your ability for mechanical weed control. That said, I've often found that transplanting by hand is more cost-effective than mechanical transplanting for any planting less than ¼ acre. Chapter 9 deals specifically with production efficiencies and explains this further, as well as flame weeding.

I transplant most crops that can be raised as starts, but not all. Short-season crops enable multiple plantings on the same land, but crops with an extended harvest frequently yield more sales for the effort of soil preparation, planting, and cultivation. A single kale plant may bear ten bunches in a season. That's potentially more than $20 in sales per plant—not bad for a crop that is also frost-proof. (Kale used to be on my "who cares?" list. No longer.) Parsley, basil, peppers, and tomatoes are some other longer-season crops with extended harvests that pay dividends. But such generalizations offer just broad brushstrokes. Some short-season crops are indeed very profitable, and some long-season ones are not.

The Crop Budget software included with this book's accompanying CD can be tailored to any farm's needs. Expense and sales names may be modified, along with their numbers, for different scenarios. Templates are easily duplicated. All the budgets are self-calculating. If you are new to Excel workbooks, begin by clicking on the tabs along the bottom of the screen (labeled WORKSHEET 1, WORKSHEET 2 . . . BEETS, CARROTS, et cetera) to view each sheet. Budgets can also be used as projections, for possible future changes or new crop trials. Use the following budgets as a guide for your farm. The best numbers are your own. Each farm has different systems, various costs associated with every component of the farm, and a unique blend of sales revenue. Take charge and discover what profit centers are hiding on your farm.

Table 4-1

Worksheet 1

Labor, Delivery, Farmers' Market, and Overhead Costs to Use in Calculating Crop Budgets

Labor Costs:

	Manager	Crew	Composite crew 1:3
Average hourly rate:	10.00	10.00	10.00
Employee taxes: 7.51%	0.75	0.75	0.75
Workers' comp: 8%	0.80	0.80	0.80
Nonassigned time: 10%	1.00	1.00	1.00
SEP-IRA: 25%			0.00
Labor costs/hour:	**12.55**	**12.55**	**12.55**

Labor costs are critical to calculating crop budgets. The farm's labor cost per hour is more than the employee's wage when employer taxes, workers' comp insurance, and nonproduction time (meetings, cleanup, maintenance) are added in. The SEP-IRA is an optional retirement plan, which is an added cost for certain qualifying employees (see chapter 6). If a farm manager is at a different pay rate, a composite rate per hour can be used. This worksheet assumes a ratio of 3 crew workers to 1 manager. For simplicity, all labor is paid the same rate in these crop budgets.

Delivery Costs:

	Produce	
Labor: load truck(s) and travel	25.10	@12.55/hr
Vehicle(s) cost at .40/mile	8.00	20 miles round trip
Cost for one delivery	33.10	
% of crop to total load	10%	for example
x number of trips	12	for example
Delivery cost for crop per season:	**39.72**	

Delivery costs can be determined for each trip, total trips per season, or the percentage cost of each product delivered. If a delivery contains equal amounts of carrots and beets, 50% of the delivery cost would be allotted to each crop.

Farmers' Market Costs:

		Calculate for ONE market
Labor: load truck(s)	12.55	1 hr (2 people @.5 hr each)
Labor: travel to market, set up	50.20	4 hrs (2 people)
Labor: market vending	100.40	8 hrs (2 people)
Labor: pack up, travel home, unpack, tally sales	37.65	3 hrs (2 people)
Vehicle(s) cost at .40/mile	8.00	20 miles round trip
Rental fees	30.00	per market
Amortized FM equipment	7.67	scales $1500, umbrellas $400, tables $200, signs $200 = $2300/15-year useful life/20 markets per season = $7.67 per market
Subtotal, cost for one market:	**246.47**	
# of markets where crop is sold	6	varies by crop
Total costs for # of markets	1478.82	
Crop sales/total FM sales	5%	varies by crop
Crop sales % x total market costs:	**73.94**	Enter in Crop Enterprise Budget under "Marketing Costs: Farmers' market expense"

The base cost for attending one market is constant irrespective of the amount of product sold (unless labor needs change). Gross sales at market must be higher than the cost; otherwise, you are losing money or personally subsidizing the market cost by not paying yourself the going labor rate. Sales need to be high enough to justify the cost of vending at market. If they are not, strive for higher sales or pursue alternative selling venues, such as CSA programs or wholesale accounts.

The total expense for equipment needed at market is amortized over the useful life of the equipment and prorated for each market. As with delivery costs above, a percentage of farmers' market expense can be assigned to different crops. The important message regarding farmers' market costs, though, is that each market costs a certain amount to attend, and that farmers' market sales must justify that expense.

Overhead Costs (annual)

Overhead costs are ones not accounted for in delivery costs, farmers' market costs, greenhouses, tractors, implement, or irrigation costs. Overhead costs are spread out over the entire farm operation and prorated to each crop or enterprise. In these worksheets, 75% of overhead expenses are apportioned to the 5 acres in cultivation, 12.5% to the bedding-plant greenhouse, and 12.5% to the in-ground tomato greenhouse. Allotment of overhead costs is somewhat subjective, but all overhead costs must be assigned. Overhead expenses allotted to the cultivated 5 acres is further broken down to overhead expense per two 350'-long beds, the equivalent of 1/10 acre.

Mortgage annual payment	600.00	farm % of total bill. Does not include house and house site portion.
Depreciation	2000.00	to account for replacement costs, excluding machinery in Worksheet 4
Property taxes	800.00	farm %
Insurance	4000.00	$3000 health, $1000 fire; not vehicle or workers' comp.
Office	1100.00	supplies, postage, subscriptions
Website	400.00	$20/month plus fees and maintenance
Travel/conferences	300.00	
Professional services	700.00	CPA, organic certification, snowplowing
Electric	600.00	farm %, w/o greenhouse electrical use
Landfill	250.00	
Telephone	550.00	farm %
Advertising	200.00	
Shop supplies, misc. repairs	500.00	tractor, implement, irrigation repairs already accounted for in Worksheet 4
Labor: management	3263.00	average 5 hrs/week, 260 hrs/year; annual labor for overseeing farm operation
Labor: office	3263.00	average 5 hrs/week, 260 hrs/year; annual labor for office duties
Labor: maintenance	653.00	average 1 hr/week, 52 hrs/year; annual labor for nonassigned maintenance work
Total overhead costs:	**19179.00**	**Allocation:** GH seedlings $2397, GH tomatoes $2397, 5A (100 beds) $14,385 = $144 per bed
Overhead per two 350' beds:	**288.00**	Per two 350' beds, for 5A (100 beds) planted to row crops. Enter on line 69 on Crop Enterprise Budget.
Overhead per greenhouse:	**2397.00**	Per 21' x 96' hoophouse: one for bedding plants, one for greenhouse tomatoes

Table 4-2

Worksheet 2

Greenhouse Flat Costs for Calculating Worksheet 3 Bedding-Plant Cost

Costs of Soil, Plastic Containers, and Labor Filling

In order to calculate what a farm-raised seedling costs, we first need to know the cost of the plastic container, the cost of the soil in the container, and the cost of labor to fill the container. Below is a table that lists common pack sizes used in greenhouse production and the associated costs with that size. A 1020 is a 10" x 20" open plastic tray. One 1020 tray will hold eighteen 3.5" square pots. A 606 is six 6-packs sized to fit a 1020 tray. An 804 is eight 4-packs sized to fit a 1020 tray. An 806 is a eight 6-packs sized to fit a 1020 tray. 128 and 98 stand for the number of molded individual cells in a 1020-sized tray. Reuse of plastic containers will lower costs.

	A	B	C	D: C/B	E	F	G: F/G	H: A + D + G
Container size	Single-use cost/flat	# of containers per yard of soil	Price per yard of soil	Cost of soil in container	# of flats filled per hour	Labor cost per hour	Cost of labor to fill flat	Total cost of plastic, soil, and labor (w/o 1020)
3.5" square pot (18/tray)	1.62	125	105	0.84	40	12.55	0.31	2.77
606	0.39	144	105	0.73	60	12.55	0.21	1.32
804	0.39	144	105	0.73	60	12.55	0.21	1.32
806	0.39	171	105	0.61	60	12.55	0.21	1.21
1020	0.72	100	105	1.05	60	12.55	0.21	1.98
128	0.95	216	105	0.49	60	12.55	0.21	1.64
98	0.95	216	105	0.49	60	12.55	0.21	1.64
6" pot: each pot	0.28	350	105	0.30	240	12.55	0.05	0.63

Table 4-3

Worksheet 3 Greenhouse Costs

Two types of greenhouse operations are portrayed: one for growing bedding plants and one for growing in-ground tomatoes. Both greenhouses are 21' x 96' hoop houses with two layers of plastic that are inflated. Each has a furnace, exhaust fan, intake shutters, and automatic controls. The longer-lived structure and equipment costs are totaled and divided by their useful life (20 years). Annual costs of heating fuel, electricity, and 5-year plastic covers are listed separately. Overhead expenses from Worksheet 1 (12.5% of total overhead) are added in after the annual expense subtotal. The bedding-plant greenhouse is more involved and listed first. The bedding-plant greenhouse benches hold 1000 flats (1020 size), and two flats can occupy the same bench space during the course of the bedding-plant season (one cycling of inventory). Worksheet 2 lists costs for plastic containers, soil, and the labor to fill the containers, as shown under *Production costs per flat*. Other production costs per flat are listed, with optional categories like thinning and fertilizing left blank for simplicity. The total cost per flat is a very useful number and will be used in the Crop Enterprise Budgets when crops are raised from transplants.

Bedding Plants, March 1st Start-up

Structure cost: 21' x 96', 2-layer poly-covered hoop house	
Frame cost $2400, installation $1004 (80 hrs), wood $300	3704.00
Furnace $2000, fans $800, installation $377 (30 hrs)	3177.00
Benches $500, plumbing $400, irrigation $400	1300.00
Total structure cost	8181.00
divide by # years of useful life	20
Annual structure cost	**409.05**
Other annual expenses:	
Poly cost $600, installation $100 (8 hrs), /5 years	140.00
Electricity 5 x $15/month	75.00
Fuel for heat 300 gallons @ $3/gallon	900.00
Watering labor 2 hrs x 50 times = 100 hrs	1255.00
Subtotal annual expenses	2370.00
Farm overhead allocation from Worksheet 1	2397.00
***Total annual expenses** with overhead allotment:*	**5176.05**

Greenhouse 1020 capacity: 1000 x 2	2000	one cycling of bench space	
Total annual expense/total flats =	2.59	per flat	
Greenhouse annual cost/flat:	2.59	2.59	2.59
Production costs per flat:	***804s***	***3.5" sq. pots***	***128s***
Cost of plastic flat, soil, labor filling	1.32	2.77	1.64
Cost of seed in flat	1.00	1.00	1.00
Labor to seed flat: 12 flats/hr = $1.05/flat	1.05	1.05	1.05
If needed: subtotal/# of finished trays			
Labor: transplant to one flat: 10 flats/hr = $1.26			
2nd plastic flat, soil, labor filling			
Subtotal for transplanted flat			
Labor moving: 60 flats/hr = $0.21/flat each move	0.21	0.21	0.21
Labor to thin: 100 flats/hr = $0.13/flat			
Fertilizer cost: $0.02/flat			
Fertilizer labor: $0.05/flat			
Total cost per flat:	6.17	7.62	6.49

Greenhouse Tomatoes, Transplanted in Ground April 1 in Northern U.S.

The annual structure cost and other annual expenses are similar to those of the bedding-plant greenhouse shown above. Overhead costs from Worksheet 1 (12.5% of total overhead) are added in after total annual expenses. This greenhouse is used to grow tomatoes in the ground for an early and extended harvest of top-quality fruit. Tomato plants are transplanted from 3.5" pots into the greenhouse soil around April 1st. Plants are irrigated with drip lines on a battery-operated water timer. The ground is mulched to reduce weeding labor. Heating and venting are on thermostatic controls. Roll-up sidewalls promote airflow when outside temperatures permit. Tomato plants are trellised from strings hanging from the greenhouse frame. A separate crop budget is calculated for greenhouse tomatoes, shown in the Crop Enterprise Budget section. The total annual expense seen below will be used as an expense in the Crop Enterprise Budget.

Structure cost: 21' x 96' two-layer poly-covered hoop house	
Frame cost $2400, installation $1004, wood $300	3704.00
Furnace $2000, fans $800, installation $377 (30 hrs)	3177.00
Total structure cost	6881.00
Annual structure cost: divide by 20 years	344.05

Other annual expenses:	
Poly cost $600, installation $100 (8 hrs), /5 years	140.00
Electricity 6 x $15/month	90.00
Fuel for heat 200 gallons @ $3/gallon	600.00
Subtotal annual expenses	830.00
Farm overhead allocation from Worksheet 1	2397.00

Table 4-4

Worksheet 4 Tractor, Implement, and Irrigation Costs

Tractor Costs

The hourly cost of a tractor is calculated by first dividing the purchase price of the tractor by the tractor's years of useful life. Next, annual expenses for repairs and fuel are added in, giving you the total cost to own and operate the tractor per year. Divide this total annual cost by the number of hours the tractor runs in a year, and the result is an average cost per tractor hour. I was surprised at first at how inexpensive running a tractor can be, but remember, a tractor used 50 hours per year has a much higher hourly rate than a tractor used 300 hours per year. The three tractors shown below are ones that I have owned, and the numbers are based on personal experience. Annual repairs are listed as an average: some years are expensive, some are not.

Tractor model	JD 2240	Ford 4000	Cub	
Original cost/useful life	*7000/25*	*4400/25*	*1000/25*	
Annual cost, w/o interest	280.00	176.00	40.00	
Average annual repairs	500.00	300.00	200.00	some years $0, some lots
Annual fuel cost @ $3/gallon	480.00	480.00	80.00	
Total annual cost	1260.00	956.00	320.00	
Hours used annually	200	300	60	
Tractor cost/hour	6.30	3.19	5.33	
Tractor driver hourly rate	12.55	12.55	12.55	
Tractor with driver: $/hour	18.85	15.74	17.88	

Implement Costs

Tracking various implements' costs is similar to tracking costs of tractors but without the fuel expense. Some implements have lots of moving parts (e.g., combines, manure spreaders) and cost more to operate than implements like a bedlifter, which has no moving parts. I list three of the more common and costly implements to run. Because a farm may have numerous implements, I make a note below these three implement costs for easy calculations to use as a shortcut for budget work.

	PTOTiller	Manure Spreader	Brush Hog
Original cost/useful life	*800/25*	*1100/20*	*600/20*
Annual cost, w/o interest	32.00	55.00	30.00
Implement annual repairs, average	20.00	20.00	20.00
Annual hours used	40	20	50
Implement cost/hour	1.30	3.75	1.00

A $500 simpler implement with a useful life of 25 years costs about $20/year to own. Figure $.50/hour for quick calculating.
A $1000 simpler implement with a useful life of 25 years costs about $40/year to own. Figure $1/hour for quick calculating.

Irrigation Costs

Irrigation costs take into account the annual equipment cost and any repair expense (similar to tractors and implements) and also time for setting up, running, and taking down (or moving) the system, calculated for the area that is watered each time. The example below shows an irrigation system that waters an acre in area and is used four times per season. The irrigation cost per acre is then calculated for 1/10 of an acre, or two 350'-long beds.

Cost of pipe, pump, sprinklers	4600.00	used PTO (power take-off) pump, 4" and 2" aluminum pipe for 1 acre
Useful life in years	25	
Annual equipment cost	184.00	
Average annual repairs	50.00	say $250 every 5 years
Total annual cost	234.00	
Total annual cost/uses per season	58.50	4 uses per season
Setup, takedown labor per irrigation area	75.30	1A coverage, 6 hrs total @ $12.55/hr
4 hours tractor use	25.20	at $6.30/hr, tractor only
Irrigation costs/irrigated area, each use	159.00	per acre
Irrigation costs for two 350' beds, each use	15.90	$7.53 labor, $8.37 machinery

Table 4-5

Crop Enterprise Budget

Crop Year:		Crop: (and specify: early, mid, late)	Broccoli
Unit Area:	Two 350' beds	Bed feet or acres:	700' or 1/10A
Today's Date:		Rows per bed & plant spacing:	2 rows/bed, 12" transplant spacing
Field:			

Note: Twenty 350' beds = 1 acre

Costs in $: Remember to prorate to unit area

NOTES: Labor at $12.55/hr. See Worksheet 1. Figures below are for two 350' beds.

	$ Labor cost	$ Machinery cost	$ Product cost	Notes
Prepare Soil:				
Disk 1x	1.26	0.73		1A at a time: 1 hr total for 20 beds = 6 mins/2 beds; $1.26L, $0.63 + .10 = $0.73M w/ JD 2240; see Worksheet 4
Chisel 1x	2.51	0.74		.5A at a time:1 hr total for 10 beds = 12 mins/2 beds; $2.51L, $0.64 +.10 = $0.74M w/ Ford 4000; see Worksheet 4
Rototill 1x, 2x				.5A at a time: 2 hrs total for 10 beds = 24 mins/2 beds; $5.02L, $1.28 tractor + .52 tiller = $1.80M w/ Ford 4000
Bedform 2x	5.02	1.48		.5A at a time: 1 hr total for 10 beds = 12 mins/2 beds; $2.51L, $0.64 +.10 = $0.74M for ONE pass w/ Ford 4000
Fertilizer	1.26	0.68	10.00	500 lbs 4-3-3/A at a time: 1 hour total for 20 beds = 6 mins/2 beds; $.1.26L, $0.63 +.05 = $0.68M, $10Pr w/ JD 2240
Manure, compost	2.52	1.02	25.00	1A at a time: compost at $25/yd, 10 yds/A; 2 hrs total for 20 beds = 12 mins per 2 beds; $2.51L, $1.26 + .75 = $2.01M, $25Pr w/ JD 2240
Other				
Plastic mulch				.5A at a time: 1.5 hr/A laying = 10 mins/2 beds; $2.09L, $0.53 +.17 = $0.70M, $20Pr w/ Ford 4000
Seed/Transplant:				
Seeding in field				2 beds at a time: 30 mins/2 beds total = $6.28L
Cost of transplants			84.00	$6.49/128 = $0.06/plant; 1400 plants at $0.06
Transplanting labor	25.23			3 rows by hand: 3 hrs/2 beds total = $37.65L; only 2 rows/bed 2 rows w/ transplanter, 6 beds at a time, 1 hr prep plants, 1.5hr x 3 people transplanting, 2 hrs machinery for 2 beds = $22.78L, $2.11 + .66 = $2.77M
Cultivation:				
Reemay on/off				For 2 beds: $105/3 uses = $35Pr, .75 hr laying = $9.41L
Hoeing 1x, 2x, 3x	12.55			at $12.55/hr: average 1 hr/2 beds; $12.55/2 beds
Hand weeding 1	25.10			at $12.55/hr: average 8 hrs/2 beds; $100.40/2 beds
Hand weeding 2				at $12.55/hr: average 4 hrs/2 beds; $50.20/2 beds
Hand weeding 3				at $12.55/hr: average 2 hrs/2 beds; $25.10/2 beds
Straw mulch				40 bales at $3, 1 hr/2 beds; $12.55L, $120.00Pr
Irrigating 1x	7.53	8.37		$7.53L, $8.37M per 2 beds, each use, w/ JD 2240
Tractor cultivating 6x	7.56	3.48		1A at a time: 1 hour/A = 6 mins/2 beds; $1.26L, $0.53 +.05 = $0.58M per pass w/ Cub mostly
Side-dressing				Spin 500 lbs 4-3-3/A, 1 hr total/20 beds = 6 mins/2 beds; $1.26L, $0.32 +.05 = $0.37M, $10Pr w/ Ford 4000
Spraying	2.51	0.74	6.00	1 hr/.5A total time = 12 mins/2 beds; $2.51L, $0.64 +.10 = $0.74M, $6Pr w/ Ford 4000
Flame weeding				10 beds/hr = 12 mins/2 beds; $2.51L, $0.64 +.10 = $0.74M, $6Pr w/ Ford 4000
Other				
Pre-harvest Subtotal:	93.05	17.24	125.00	= 235.29 Pre-harvest cost for two beds

Harvest:

Total yield for two 350' beds = 36 cases — season average: 500 bunches, 14-count case

Total hours to harvest two 350' beds: 6 hrs — 6 cases/hour

	Labor cost	Machinery cost	Product cost	Notes
Field to pack house	75.30			at $12.55/hr; 6 hrs
Pack house to cooler	37.65			at $12.55/hr; 12 cases/hour packing
Bags, boxes, labels			19.44	$0.25/bag, $1.00/box, $0.07/label; $1.07 per box/ 2 uses
Delivery	30.12	9.60		See Worksheet 1.
Post Harvest:				
Mow crop	2.09	0.70		6 beds at a time: 10 mins/2 beds; $2.09L, $0.53 +.17 = $0.70M w/ Ford 4000
Remove mulch				1 hour/2 beds: $12.55L
Disk	1.26	0.73		$1.26L, $0.63 +.10 = $0.73M w/ JD 2240, see disking above.
Sow cover crop: spinner	1.26	0.68	8.00	1A at a time: 1 hr/20 beds = 6 mins/2 beds; $1.26L, $0.63 + .05 = $0.68M, $8Pr w/ JD 2240
Sow cover crop: Brillion				1A at a time: 2 hrs/20 beds = 12 mins/2 beds; $2.51L, $1.26 + .20 = $1.46M, 8Pr w/ JD 2240
Other				
Post-harvest Subtotal:	240.73	28.95	152.44	= 422.12 Harvested cost for 2 beds

	Labor cost	Machinery cost	Product cost	Notes
Marketing Costs:				
Labor: sales calls for season (for this crop only)	6.28			Average 10 mins/week for 3 weeks: .5 hr
Commissions				Commissions, if any, to growers' co-op, broker, or salesperson
Farmers' market expense	60.24	4.70	9.00	See Worksheet 1.
Total Crop Costs:	307.25	33.65	161.44	= 502.34 Total crop costs
Overhead Costs:	288.00			Apportionment for two 350' beds, see Worksheet 1.
Total Costs: Crop & Overhead Total:	790.34			Total costs per two 350' beds

Sales:	# of units	Price per unit	Total $
Retail:	12.00	31.50	378.00
Wholesale:	24.00	22.00	528.00
Other:			0.00
Total units	36.00		
Total Sales:			906.00 For two 350' beds

Net Profit: Total sales – total costs =	115.66	**Net profit for two 350' beds (1/10 acre)**
Net Profit/Acre:	1156.60	Standardize to one acre
Cost/Unit:	21.95	Total cost/total units
Net Profit/Unit:	3.21	Net profit/total units

NOTES:

Table 4-6

Crop Enterprise Budget

Crop Year:		Crop: (and specify: early, mid, late)	Kale: bunches	Unit Area:	Two 350' beds
				Bed feet or acres:	700' or 1/10A
Today's Date:		Rows per bed & plant spacing:	2 rows/bed, 24" spacing, transplanted		
Costs in $:		Remember to prorate to unit area		Field:	

Note: Twenty 350' beds = 1 acre

NOTES: Labor at $12.55/hr. See Worksheet 1. Figures below are for two 350' beds.

	Labor cost $	Machinery cost $	Product cost $	Notes
Prepare Soil:				
Disk 1x	1.26	0.73		1A at a time: 1 hr total for 20 beds = 6 mins/2 beds; $1.26L, $0.63 + .10 = $0.73M w/ JD 2240; see Worksheet 4
Chisel 1x	2.51	0.74		.5A at a time:1 hr total for 10 beds = 12 mins/2 beds; $2.51L, $0.64 +.10 = $0.74M w/ Ford 4000; see Worksheet 4
Rototill 1x, 2x				.5A at a time: 2 hrs total for 10 beds = 24 mins/2 beds; $5.02L, $1.28 tractor + .52 tiller = $1.80M w/ Ford 4000
Bedform 2x	5.02	1.48		.5A at a time: 1 hr total for 10 beds = 12 mins/2 beds; $2.51L, $0.64 +.10 = $0.74M for ONE pass w/ Ford 4000
Fertilizer	1.26	0.68	10.00	500 lbs 4-3-3/A at a time: 1 hour total for 20 beds = 6 mins/2 beds; $.1.26L, $0.63 +.05 = $0.68M, $10Pr w/ JD 2240
Manure, compost	2.52	1.02	25.00	1A at a time: compost at $25/yd, 10 yds/A; 2 hrs total for 20 beds = 12 mins per 2 beds; $2.51L, $1.26 + .75 = $2.01M, $25Pr w/ JD 2240
Other				
Plastic mulch				.5A at a time: 1.5 hr/A laying = 10 mins/2 beds; $2.09L, $0.53 +.17 = $0.70M, $20Pr w/ Ford 4000
Seed/Transplant:				
Seeding in field				2 beds at a time: 30 mins/2 beds total = $6.28L
Cost of transplants			42.00	$6.49/128 = $0.06/plant 700 plants
Transplanting labor	25.23			3 rows by hand: 3 hrs/2 beds total = $37.65L 2/3 of 3-row time
				2 rows w/ transplanter, 6 beds at a time; 1 hr prep plants, 1.5hr x 3 people transplanting, 2 hrs machinery for 2 beds = $22.78L, $2.11 + .66 = $2,77M
Cultivation:				
Reemay on/off				For 2 beds: $105/3 uses = $35Pr, .75 hr laying = $9.41L
Hoeing 1x, 2x, 3x	25.10			at $12.55/hr: average 1 hr/2 beds $12.55/2 beds
Hand weeding 1	50.20			at $12.55/hr: average 8 hrs/2 beds $100.40/2 beds
Hand weeding 2	25.10			at $12.55/hr: average 4 hrs/2 beds $50.20/2 beds
Hand weeding 3				at $12.55/hr: average 2 hrs/2 beds $25.10/2 beds
Straw mulch				40 bales at $3, 1 hr/2 beds; $12.55L, $120.00Pr
Irrigating 1x	7.53	8.37		$7.53L, $8.37M per 2 beds, each use, w/ JD 2240
Tractor cultivating 6x	7.56	3.48		1A at a time: 1 hour/A = 6 mins/2 beds; $1.26L, $0.53 +.05 = $0.58M per pass w/ Cub mostly
Side-dressing				Spin 500 lbs 4-3-3/A, 1 hr total/20 beds = 6 mins/2 beds; $1.26L, $0.32 +.05 = $0.37M, $10Pr w/ Ford 4000
Spraying	5.02	1.48	12.00	1 hr/.5A total time = 12 mins/2 beds; $2.51L, $0.64 +.10 = $0.74M, $6Pr w/ Ford 4000
Flame weeding				10 beds/hr = 12 mins/2 beds; $2.51L, $0.64 +.10 = $0.74M, $6Pr w/ Ford 4000
Other				
Pre-harvest Subtotal:	158.31	17.98	89.00	= 265.29 Pre-harvest cost for two beds

Harvest:

Total yield for two 350' beds = 2800 bunches

Total hours to harvest two 350' beds 18.7 hrs 150 bunches/hr

	Labor cost $	Machinery cost $	Product cost $	Notes
Field to pack house	234.69			at $12.55/hr 18.7 hrs
Pack house to cooler	292.42			at $12.55/hr 120 bunches/hr, 23.3 hrs
Bags, boxes, labels			166.92	$0.25/bag, $1.00/box, $0.07/label 156 18-count boxes at $1.07
Delivery	30.12	9.60		See Worksheet 1.
Post Harvest:				
Mow crop	2.09	0.70		6 beds at a time: 10 mins/2 beds; $2.09L, $0.53 +.17 = $0.70M w/ Ford 4000
Remove mulch				1 hour/2 beds: $12.55L
Disk	1.26	0.73		$1.26L, $0.63 +.10 = $0.73M w/ JD 2240, see disking above.
Sow cover crop: spinner	1.26	0.68	8.00	1A at a time: 1 hr/20 beds = 6 mins/2 beds; $1.26L, $0.63 + .05 = $0.68M, $8Pr w/ JD 2240
Sow cover crop: Brillion				1A at a time: 2 hrs/20 beds = 12 mins/2 beds; $2.51L, $1.26 + .20 = $1.46M, 8Pr w/ JD 2240
Other				
Post-harvest Subtotal:	720.15	29.69	263.92	= 1013.76 Harvested cost for 2 beds

	Labor cost $	Machinery cost $	Product cost $	Notes
Marketing Costs:				
Labor: sales calls for season (for this crop only)	6.28			Average 10 mins/week for 3 weeks: .5 hr
Commissions				Commissions, if any, to growers' co-op, broker, or salesperson
Farmers' market expense	60.24	4.70	9.00	See Worksheet 1.
Total Crop Costs:	786.67	34.39	272.92	= 1093.98 Total crop costs
Overhead Costs:	288.00			Apportionment for two 350' beds, see Worksheet 1.
Total Costs: Crop & Overhead Total:	1381.98			Total costs per two 350' beds

Sales:	# of units	Price per unit	Total $	
Retail:	460.00	2.00	920.00	
Wholesale:	2340.00	1.25	2925.00	
Other:			0.00	
Total units	2800.00			
Total Sales:			3845.00	For two 350' beds

Net Profit: Total sales – total costs =	2463.02	**Net profit for two 350' beds (1/10 acre)**
Net Profit/Acre:	24630.20	Standardize to one acre
Cost/Unit:	0.49	Total cost/total units
Net Profit/Unit:	0.88	Net profit/total units

NOTES:

Table 4-7

Crop Enterprise Budget

Crop Year:

Crop: **Tomatoes: greenhouse**
and specify: early, mid, late

Unit Area: **21' x 96' greenhouse**
Bed feet or acres:

Today's Date:

Rows per bed & plant spacing: 5 rows total, 12" in row spacing, non-grafted plants

Costs in $:

Remember to prorate to unit area

Field:

	$ Labor cost	$ Machinery cost	$ Product cost	NOTES: Labor at $12.55/hr. See Worksheet 1.	
Prepare Soil:					
Spread fertilizers, compost	25.10		120.00	4 yards compost, 50 lbs fertilizer	
Rototill	12.55	3.00		1 hr	
Rake, handwork	25.10			2 hrs	
Set drip lines, patch, check	25.10		25.00	2 hrs	$40 drip lines/2 uses plus fittings
Install mulch and anchor	12.55		28.00	1 hr	$200 weed mat/10 yds, anchors
Tighten greenhouse, other	25.10			2 hrs	
Heat, vent, alarm ready	25.10			2 hrs	
Other					
Seed/Transplant:					
Cost of transplants			243.00	450 plants needed/greenhouse, $0.54/3.5" pot	
Transplanting labor	50.20			4 hrs	
Cultivation:					
Drop strings	25.10		5.00	2 hrs	
Clip strings	25.10			2 hrs	
Prune and trellis 7x	329.44			Average: .75 hr/row, 3.75 hrs each time = 26.25 hrs total	
Weed holes, edges 3x	75.30			6 hrs total	
Prune leaves, sweep up 3x	112.95			9 hrs total	
Top plants 9/1	37.65			3 hrs total	
Roll up and down sides	58.99			4 mins/time x 70 days = 4.7 hrs	
Pre-harvest subtotal:	865.33	3.00	421.00	= 1289.33	Pre-harvest cost
Harvest:	Total yield for greenhouse =			300 15-lb boxes	at 10 lbs marketable fruit/plant
	Total hours to harvest greenhouse			60 hrs	average: five 15-lb boxes/hr
Field to pack house	753.00			at $12.55/hr	60 hrs
Pack house to dock	376.50			at $12.55/hr	at 10 boxes/hr sorting and folding up boxes
Bags, boxes, labels			321.00	$1.00/box, $0.07/label	300 at $1.07
Delivery	30.12	9.60		See Worksheet 1.	
Post Harvest:					
Detrellis and remove plants	75.30			6 hrs total	
Sweep and fold mulch	12.55			1 hr	
Move drip lines	12.55			1 hr	
Post-harvest subtotal:	2125.35	12.60	742.00	= 2879.95	Harvested cost for greenhouse
Marketing Costs:					
Labor: sales calls for season (for this crop only)	25.10			Average 10 mins/week for 12 weeks = 2 hrs	
Commissions				Commissions, if any, to growers' co-op, broker, or salesperson	
Farmers' market expense	60.24	4.70	9.00	See Worksheet 1.	
Total Crop Costs:	2210.69	17.30	751.00	= 2978.99	Total crop costs
Greenhouse & Overhead Costs:	3227.00			Greenhouse annual expenses: $830; greenhouse overhead allocation: $2397. See Worksheet 3.	
Total Costs: Crop & Overhead Total:	6205.99				

Sales:	# of units	Price per unit	Total $
Retail:	100.00	48.75	4875.00
Wholesale:	200.00	36.00	7200.00
Other:			0.00
Total units	300.00		
Total Sales:			12075.00

Net Profit: Total sales – total costs =	5869.01	**Net profit for Greenhouse**
Net Profit/Acre:		Not applicable
Cost/Unit:	20.69	Total cost/total units
Net Profit/Unit:	19.56	Net profit/total units

NOTES:

– 5 –
Marketing Strategies

Business guru Peter Drucker once stated, "The purpose of business is to create and keep a customer." I say this only after subjecting you, the reader, to four full chapters on the complexities of how to make a profit for your farm business. Oops. Maybe it is too obvious that a business has to remain profitable. After all, the essence of any business is to generate a positive net return. But Drucker's comment underlines the importance of the marketplace to business survival. No business can make a profit without first garnering some income. And that income comes from new customers, and repeat customers. Or, said more somberly: no customers, no income, no business.

The Marketing Circle

Successful businesses, whether small farms or giant corporations, follow an ongoing practice of market analysis. A concept called the Marketing Circle visually depicts this practice. With no beginning and no end, the Marketing Circle asks four questions to help you better serve your marketplace.

Begin by brainstorming ideas about each question. *How will the market know I have what it needs?* I can knock on doors, make phone calls, advertise, post flyers, generate free press, give away samples, use a website and blog, or drop leaflets from an airplane. The list can go on and on. Sort through your brainstorm list and rank each item by importance. Make a plan: which ideas to do when, and how much they will cost.

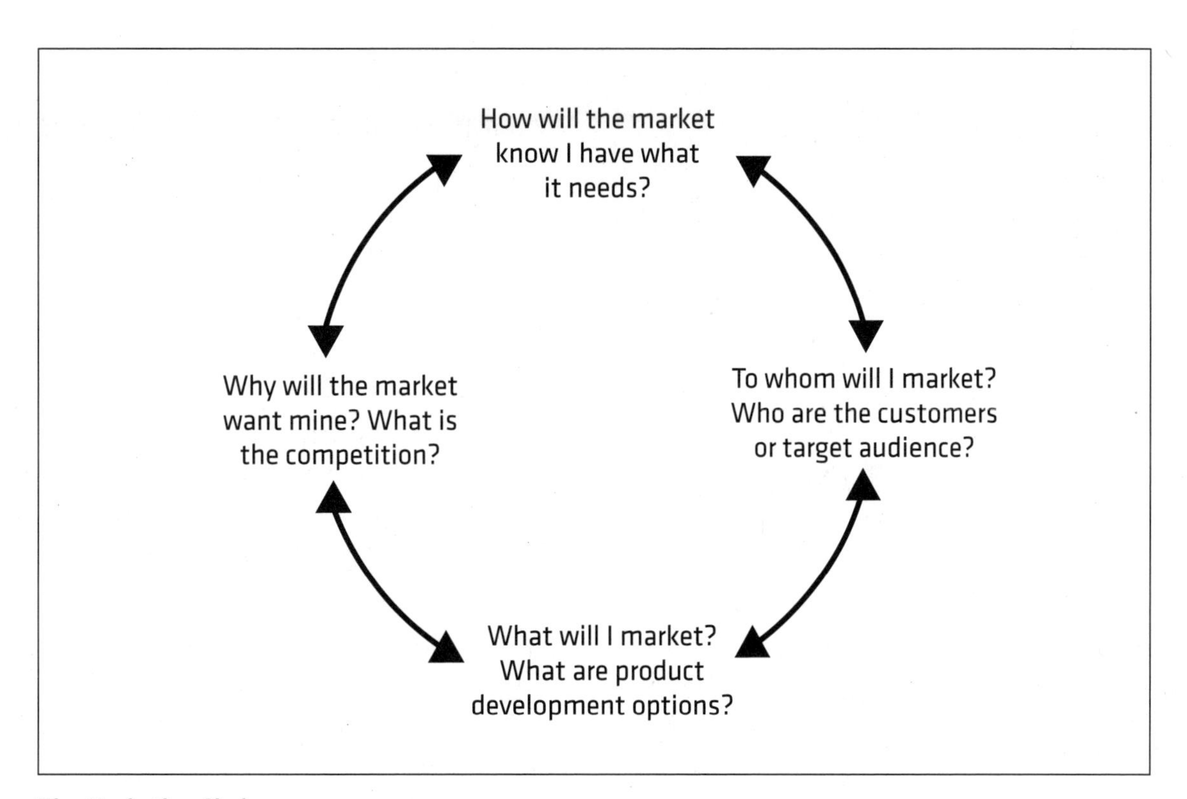

The Marketing Circle.

To whom will I market? Again, brainstorm. Imagine all the possibilities. Retail, wholesale? Stores, restaurants, farmers' markets, CSA, farm stand, pick-your-own, growers' co-op? Which restaurants or stores to choose? What geographic area? Who is the best match for my product? Who is my typical customer? Who would be good as a business ally? A practical way to start is to take your ideas and start knocking on doors (sometimes the back door). Face-to-face interactions carry weight. State your business to the potential buyers and ask if they are interested in your product. Ask questions. Whom do they buy from now? Would they be interested in buying from you? What are customary prices? What's the volume purchased each week? What would the buyer like to see for product, communication, and delivery? How can *you* be of service? As a relationship develops, make sure to follow through. If you visited in the off-season, touch base again as the new growing season approaches, and again midseason. Start off small; don't bet the farm on an unproven relationship.

Why will the market want my product? Assess the competition. Place yourself in the buyer's shoes. Why would the buyer want to buy from someone else? Different or superior product? Better price? More service? Friendlier? Why should a store's produce buyer or a restaurant chef risk changing habits and suppliers to support you? And why would a customer at a farmers' market stop at your stand and not at another?

Years ago I studied at a language school in Cuernavaca, Mexico, and shopped at the local farmers' market. As I entered this market for the first time, four similar-looking pickup trucks filled with sweet corn were lined up on one side, selling off the tailgate. The first pickup truck had a guy sitting on the tailgate smoking a cigarette, looking down. The next truck had a radio blaring and no one to be found. A couple (presumably husband and wife) were arguing about something at the third truck. I passed the fourth truck and was greeted: "Good morning! Can I help you?" It was clear who would get my business. The sweet corn on all four trucks looked to be the same quality, but the vendor from the last truck earned my patronage. What's more, he would have my business for as long as I was in the area. He *got me*—as a customer—and I would keep coming back to him again and again, unless he did something to offend me.

The remaining question—*What will I market?*—opens the door to inventiveness and potential. An entrepreneur running a business is like an artist with a blank canvas. Any direction is possible; you call the shots. It is a highly creative process. What do you like to do? What are the possibilities? If you have already studied the preceding four chapters, you are armed with meaningful information to answer these questions (and I'm off the hook with Peter Drucker's opening line that business's primary purpose is not profit). Which products do I want to sell *that are also profitable*? Now, there's a shortcut you can bank on. Income can be generated on the farm by any number of crops, animals, or their products. How can I expand my market for the biggest net return enterprises? The choices are yours.

The Marketing Chart found in chapter 2 succinctly depicts your projected crop selection, crop amounts, and where each will be sold. The gross sales generated in this projection are necessary to achieve your desired net profit. Use this one-page picture of your farm business when asking the questions listed above. Do other accounts or crops deserve exploration? Are all existing crops and accounts still desired? Think outside the box. Take time to consider the possibilities, and enjoy the process.

Rejection 101

Farmers aren't always the best salespeople. Neither are a lot of other people. A common stumbling block trips up the best-laid marketing plans. This block is fear, and it can prevent us from being exceptional salespeople. But selling product is an integral part of being in business.

Bumper crops are great if you can sell them. When I'm swimming in beautiful red, ripe tomatoes and see plants loaded with more fruits ready to pick, I need to look outside my normal market channels, and fast. I survey my options: stores and restaurants that are not regular accounts but that I have done business with in the past;

wholesalers; and neighboring farm stands and CSAs. I get out the phone book and make a list of potential tomato buyers to call. But then something happens as I reach for the phone . . .

Have you ever found it hard to pick up the phone and call ten new potential buyers? I have, because generally I get only one yes for the ten calls. That's nine no responses for every yes. Not a great average. But so what? What is the big deal? Intellectually and rationally, I understand the reality that the potential buyer already has enough product, or already has a satisfactory supplier. Ultimately, though, my behavior is irrational: It is resistance to *rejection*—the fundamental and unconscious feeling of dismissal and repudiation. Your psyche cannot separate this connection all the time. On some elemental level, the phone may trigger feelings of a potential high school prom date saying no, or, deeper still, rejection from your parents as a kid.

Good salespeople seem to have a thick skin, being able to take rejection repeatedly, and not take it personally. That's the key—*not taking it personally*. Of the nine no calls, *consciously* I understand that the potential buyer just doesn't need or want what I'm trying to sell. If someone at the time called me up and asked if I wanted to buy tomatoes, my answer would be an emphatic (and transparently so) no. The solution to the dilemma is to realize *deep down* that rejection in sales is not the same as a traumatic rejection of your past. No one likes being rejected, but we must remember to separate marketplace rejection and personal rejection, because sometimes the reality is *twenty* no responses for each yes. Welcome to Psychoanalysis 101. (I'm sorry, your hour is up. That'll be sixty bucks.)

The only difference between success and failure is that people who succeed are those who get back up again after falling down. And if you need more motivation, remember Vince Lombardi's revelation: The only place where success comes before work is in the dictionary. Go pick up the phone.

A Word on Pricing

Produce prices are defined by retail and wholesale markets and may be fairly stable from year to year. At the time of this writing, wholesale prices for twenty-four-count cases of lettuce in season are roughly $18/case by the pallet (growers' co-op to large wholesale accounts), $28/case to local wholesale (stores, restaurants, food co-ops), or $48/case retail (farm stand, CSA, farmers' market). Retail customers are accustomed to paying $1.75 to $2.50 per head of lettuce. The range of lettuce prices reflects, in part, the amount of marketing needed to sell that case. Setting up and staffing a farmers' market booth costs more than a quick phone call and delivery to a wholesale account.

Your job as farmer is to grow a case of lettuce for less than the price you receive for that case. If you produce a case of lettuce for 90 percent of the sales price, fine. If you can produce it for only 30 percent, so much the better. Should you pass on the additional savings to your buyer? That is your choice, but generally a retailer will realize the difference in profit, not you. I recommend pocketing those profits yourself. That may seem to be money grubbing or unfair, but given the vagaries of farming, like weather, you will need those profit centers to offset less fortunate production outcomes. Profit is not a four-letter word, though a lot of people feel that it is. You should feel good about making some money for all the time spent, all the trial-and-error experiences, and all the risks taken. Defy the paradigm of Poor Dumb Farmer. Let me reiterate my advice from chapter 4 on highly profitable crops. It is counterintuitive to raise the sales price on an already profitable enterprise—there is no inherent need to. But why not? If the marketplace supports a higher price, all the better for you. Business will cycle up and down, so make hay while the sun shines.

Brand ID

When I first heard the term *brand ID* I thought it was cold, corporate, and reserved for

Signs at market.

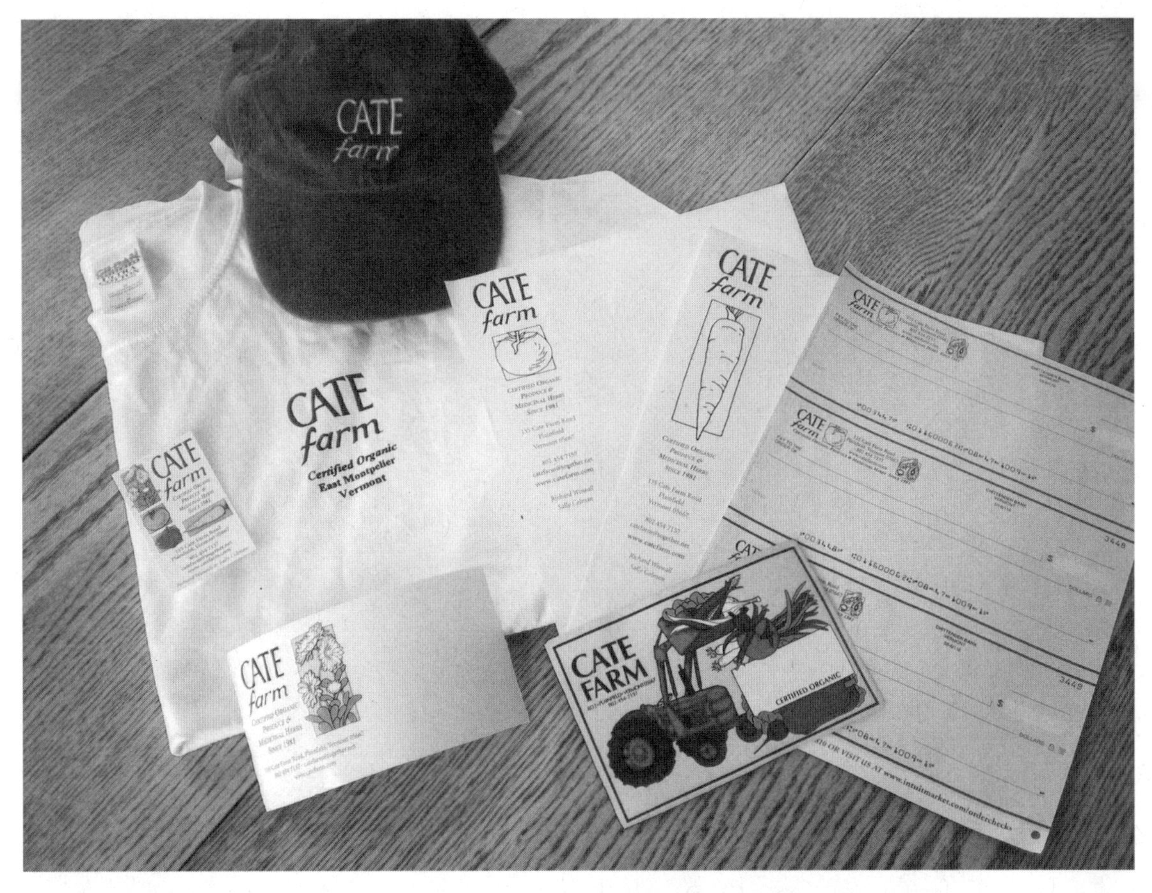

Brand ID on different items.

multinational businesses. Why should a farm care about such matters?

When you see a swoosh logo, you immediately think of the company Nike. Even the word *swoosh* exudes Nike-ness. Recognizable from afar, it is simple and packed with identity. Coca-Cola's elementary red-and-white logo is a potent conveyor of the product. Thirsty for a soda? Head to the red-and-white design, no lettering needed.

Symbols are shortcuts to convey meaning or a message. Why not use this advantage to promote your own farm business? A unified

Magnetic truck sign.

Plastic bags with farm brand.

look makes it easier for anyone (new and old customers) to readily identify you. Creating a thematic "look" for your business isn't difficult. Many aspects of promotion are already commonplace but underutilized or not coordinated: farm invoices, farm checks, business cards, signage at farmers' market and farm stand, produce bags, case labels, farm truck lettering/artwork, T-shirts, hats, and stationery letterheads. Choosing a theme for color, style of letters, and artwork for a logo unifies all these different outward presentations of your farm business. A symbol's color, style, and shape weave a thread of unified identity—your farm.

Act Professional

When I first started out farming, I'll admit that I didn't know what I was doing. Desire to farm overcame ignorance. Some years later, even with the farm's apparent success, I still hesitated to call myself a "professional" because I realized how much I *still* didn't know. Nowadays, I do use the term *professional,* albeit jokingly. When a customer exclaims, "How did you get these carrots so early?" I respond with a smile, "Hey, we're professionals."

What makes someone a professional? I profess to say it is competence and experience in a particular specialty. But *acting professionally* doesn't necessarily fit that bill. Plenty of professionals don't act professionally, no matter what their specialty is. Assorted elected officials, sports superstars, movie celebrities, and even some plumbers come to mind (crack intended).

What do I mean by the term *acting professionally,* and what's the tie-in with farming? A grower presenting at a conference I once attended talked about finally landing a certain large account for his products. The account was very selective. Somewhat intimidated, the farmer made the initial delivery to the store's delivery dock himself. Above the door, in big letters, was a sign that read:

The Five Rules of Doing Business with Us

1. Deliver in full, and on time.
2. Deliver in full, and on time.
3. Deliver in full, and on time.
4. Deliver in full, and on time.
5. Deliver in full, and on time.

I guess some people needed a reminder to act professionally. And the message was perfectly clear.

What are other traits of acting professionally besides sticking to your commitments and punctual delivery? Punctuality in all respects is near the top of the list. Being on time for meetings, appointments, work with employees, and get-togethers shows that you care about and have respect for the other people involved. Calling in orders at the same time of the week and same time of day is helpful to the buyers—they can count on you like clockwork. The same goes for deliveries, as seen above. Ask your buyers what works best for *them,* and try to accommodate their wishes. Some negotiation may be appropriate—as is common with product prices—but your willingness to work out an answer that suits both parties can only strengthen the relationship. Remember that truly sustainable business happens only when every player on the food chain is content. That includes you *and* them.

Hand in hand with punctual behavior is responding to others. Numerous times I have phoned, left a message with someone, and never heard back. This includes people in both my personal and professional life. Seemingly insignificant, it is not always forgotten. No matter how busy someone is, common courtesy should prevail. Treat others as you would like to be treated. If you are too busy at the moment to adequately respond, simply say so—quickly reply that you will get back to them later (and do). If you find that you are receiving too many phone calls, ask why this is so. Is the local co-op referring customers to call you directly? Does your website encourage telephone calls? Can you preempt commonly asked questions like, "When are you open?" with a recording on your answering machine? When you are not a black hole of communication—that is, a one-way mailbox—people notice, and they appreciate your effort. Farmers, like most businesspeople, need to interact with the public and act professionally.

CSA Potholes

Community-supported agriculture (CSA) programs are wonderful. What better way to transact business between farmer and consumer? CSA members help the farmer by paying up front for the entire season, guaranteeing a market for the farm's products, reducing packaging, and becoming directly involved in the farm. Both farmer and consumer benefit by cutting out the middleperson.

My wife and I had a popular CSA for six years and found out through trial and error some tips for success.

Farmers' feelings around the CSA concept vary. Some people like that it skirts the normal cash transaction for each product, while others treat their CSA as a subscription service to give discounts off individual items to members. But no matter where your CSA falls on

CSA farm distibution. Credit: Vern Grubinger

this continuum, the fact remains that money is traded for farm product, even if it happens before the season begins and in one bulk payment.

Whether or not the farm highlights the fact to its CSA membership, each week's share represents a certain payment from each member. On average, if a season's CSA share costs $500 for twenty weeks, the farmer receives $25 for each week's amount of product. Using this $25/week as a guide, what happens in peak season when the harvest is abundant? Do you simply distribute all the harvest evenly to your subscribers?

The value of the weekly share is important, so: (1) Don't give away the farm, and (2) don't inundate CSA members with too much food. Generosity is a good thing, but it may backfire when members return next year purchasing only a half share (thinking they will receive just as much as this year). Members abhor wasting good farm-fresh food and respond accordingly—maybe by not re-signing at all.

If an oversupply of product causes the share value for a particular week to pile up to $50, not the $25 average that is received in payment, the farm value of the product is cut in half. For example, the lettuce that would normally fetch $2 per head now brings in only $2 *per two heads*. All items in the share would be sold at half price. Farm expenses remain the same, though, seriously eroding the overall bottom line. Wouldn't it be better for farmer and CSA member alike to keep the trade fair?

Over the course of the season, keep the value of the CSA share close to the amount of money you are receiving for it. Assign a reasonable value to each share item and monitor the total share price each week to keep on track ($25/week in the above example). Crops may be substituted for one another, or lessened or increased in volume. When a CSA farm has an overabundance of supply, other sales outlets should be explored (and a note made to adjust production for next year). Case-lot pricing may even be offered to CSA members—say, for freezing spinach or canning tomatoes.

CSA programs work well in concert with another marketing channel or two, for the ability to sell off bumper crops. That said, a *stand-alone CSA* is also a fine business model, and it is one case where the farm's overall bottom line may not need to be dissected. If CSA members support a farm budget that works for everyone involved, farmers included, individual crop profitability matters less. Don't get me wrong, the knowledge of each crop's contribution to the overall farm budget is still extremely valuable and will dramatically increase the efficiency of the CSA. But CSA whole-farm budgets for a highly diverse crop mix with a prepaid guaranteed market and supportive membership may have their place.

– 6 –

Effective Management

Have you ever sat at your desk and felt so overwhelmed with the amount of work you had to do that you became paralyzed? I sure have. My eyes glaze over and I simply look off into the distance. My wife calls this the "taxidermy stare." Needless to say, this is *not* one of the seven habits of highly effective people.

To avoid these bouts of paralysis, I follow some simple steps, and remember that *effective management is tackling the most important things first, and following through to make sure they get done.*

The Clean Desk

Let's start at the office desk. Is the top of your desk visible, or is it covered in papers? If papers dominate, then it is time for some action. Begin by going through the papers one by one and then filing them away to an appropriate file or getting rid of them by throwing them out (recycling, that is). Another possibility is to give the paper to someone else, if appropriate. The important thing is that you make a decision *now* where that paper will go. Ask yourself if the piece of paper is something you really need to keep. If it is, then what file should it reside in? Get in the habit of putting the date on every paper you handle, no matter what it is. Dates give insights later on that may influence the interpretation of the paper's contents.

Get comfortable with using a filing system. Manila folders with clear file names safely house any papers, and they make documents easy to retrieve. Why have papers shuffling around on the surface of your desk when they can be stored nearby for easy reference? Use multicolored folders with left-hand, middle, and right-hand tabs. Group together files of similar types (for example, farm, house, personal), or file them alphabetically. Alternatively, place the most commonly accessed folders near the front of the files—whatever resonates with you. Create a one-page index for the front of the filing cabinet showing all file names inside. Your goal should be to create a system that facilitates putting papers away and enables you to find them again easily.

In addition to manila folders, I keep a few pocket folders in which papers can be tucked on either side or left loose in the middle. I keep one of each in my desk drawer for:

- **Employee papers:** Past and present W-2s, I-9s, Form 943s, W-3s, payroll records, et cetera (covered later in this chapter).
- **Insurance:** Different policies in each pocket.
- **Paid bills:** One pocket for January–June, one for July–December. (More on this in chapter 7.)

Pocket folders also work well for storing papers of organizations you belong to, in case you want to bring to meetings one folder that contains separate compartments. I keep these folders on my office shelf, labeled on the outside fold with a bold marker: DEEP ROOT GROWERS' CO-OP, CERTIFICATION, LOAN FUND, and so on.

It is much easier keeping your desk clean once you develop your own filing system. If a particular paper could be filed in two or more different files, choose the most likely one. When looking for papers in your files, begin your search in the most likely file, and if it's not there, look in the next most likely. For instance, if I'm looking for a saved newspaper article on biodiesel, I might start by looking in the BIODIESEL folder. If it's not there, I'd look in my ARTICLES folder. Still no article? I'd check my HEATING folder.

There are some other pieces of paper that will end up on your desk that will need a temporary resting area on top of your desk instead of in the file drawer. These items and their respective places are listed below:

- **Unpaid invoices:** Place in an alphabetical accordion folder on or near your desk.
- **Bills to pay:** Put in an upright paper holder, or cubby, somewhere on or near your desk.
- **Credit card slips:** Like bills to pay, above, but keep in separate holder or cubby.
- **Account slips:** Transaction records of purchases you made at local businesses that are billed monthly. Place in a separate holder or cubby near bills to pay.
- **Payments for deposit:** Checks and cash to go to the bank.

Temporary resting areas along the paper trail.

Chapter 7 explains how to process these invoices, bills, and slips. The first task at hand, though, is having an assigned spot for the paper documenting these transactions.

The Master List

As you clean off your desk papers one at a time, certain papers may require you to take some sort of action, such as a bank statement that needs reconciling, or a note saying beet seed is all gone, or that the shop roof leaks. Put any of these tasks on a master to-do list. In the off-season, I use a clipboard for my list, but in the busy season I need more room and use a big piece of lecture pad paper on my office wall. I have different categories on the master list, such as office, greenhouses, repairs, farmers' market, fieldwork, miscellaneous, and long term. Anything and everything that you need to do is on that one sheet of paper. The papers associated with any given task can be filed away (if appropriate) for retrieval once you're working on that specific task.

The next step in being a highly effective

manager is to prioritize the most important tasks on the master list *and assign time to work on them on a calendar*. This is a crucial step. *Effective management comes from tackling the most important work first, and following through to make sure it gets done.* By putting tasks on the calendar, a block of time is designated to accomplishing them.

When will I reconcile the bank statement, order seed, or fix the shop roof? I look at my calendar and block out enough time on an available day to finish the job at hand. If it is a one- or two-hour job, I allow two or even three hours for it without interruption. Why answer the phone when you are right in the middle of a project? Let the answering machine take the message and return the call when you have time. Treat your time preciously and be realistic in your expectations of how long something will take to finish. If all goes well and you are ahead of schedule, you can always work on a future task.

Certain times or days on the calendar are predetermined, such as farmers' market vending days, time for picking before market, days for calling orders and picking, CSA share distribution, and meetings that are scheduled in advance. Use open parts of your weekly routine to tackle your most important tasks. Leave some open blocks to address things that pop up. I try to keep a regular schedule for office duties (usually one or two mornings per week), as well as downtime (at least Sunday afternoon). Besides important farm jobs, I also put on the calendar farm and personal responsibilities that are preset, like monthly loan payments, employee tax filings, quarterly tax payments, annual land taxes, rent, children's activities, dentist appointments, and birthdays. Be sure to assign time for important nonfarm activities, too, including personal time with your spouse and family—otherwise it is all too easy to fill every waking hour with work.

Calendars come in all shapes and sizes, from the common desktop blotter size to portable planner date books to handheld electronic devices. Any calendar will work as long as you can write tasks on individual days of the week. I prefer a large 18-by-22-inch paper desk calendar that resides on my office desk. There is plenty of room to record tasks and events, and it is centrally located. At times, I'll need my calendar information while I'm off the farm, such as when I'm in a meeting and need to schedule a future gathering. Since this desk calendar is unwieldy to carry around, I use a small datebook as a duplicate of the larger calendar. Fixed events and important tasks are written on the desk calendar and then copied to its mobile companion. The portable datebook is smaller and cannot contain as much detail as its larger desk model, but it holds enough information for remote planning purposes.

After adding to the schedule my fixed daily or weekly routine of farm duties (like farmers' market or Monday-morning desk jockey) or personal responsibilities (children's music lessons or dentist appointments), I leave some open time slots to accomplish what other things need to get done. I prioritize the tasks at hand and place the most important ones on the calendar, making sure I assign enough time to complete each job. So what is the next step?

Whenever I complete a task, I cross it off the calendar. In a perfect world, every task would be finished on schedule as planned. But the wild card of weather, along with other curveballs thrown at you, makes farm planning less than a precise science. The best-laid plans sometimes have to be postponed. If I don't manage to finish the job or even get to it in the first place—and this is very important—I *move the uncompleted task to a future date to make sure it gets done.* If I didn't repair the leaking shop roof as planned on my calendar, I'll have a rude reminder on the next rainy day. *Effective management comes from doing the most important things first, and following through to make sure they get done.*

Some long-range, but nevertheless important, projects are put off to future months, or even years. I make a note in the long-range category on the master list of when I hope to address these endeavors. I also use small sticky notes with my desk calendar for some medium- to long-range undertakings with less defined time frames. I place the notes on the calendar page when I hope to have time to work on them, or otherwise at the end of the current calendar year to carry over into the following year. If I flip a calendar page to the new month and see a previously written note, I assign a definite time slot to complete the task within that month. If I need or decide to put the project off to a later time, I move the sticky note to a future month.

Time Quadrant

How do we best decide which tasks are the most important to do? With a very long to-do list, why is it that some jobs rise to the top of the list and demand our attention?

Every day, hundreds of messages come at you that affect your decisions. Possible sources include your family, newspaper articles, the Internet, radio, your telephone, salespeople, or me. Some messages are direct and clear, like *did you fix the shop roof yet?* Others are less overt, like radio ads that get partially tuned out of your consciousness. But all messages influence what we choose to do with our time.

Stephen R. Covey, in his book *The 7 Habits of Highly Effective People,* introduces the concept of the Time Quadrant Matrix. The quadrant divides your time and tasks into four areas defined by two lines, as shown below:

Quadrant I: Urgent, Important	Quadrant II: Not Urgent, Important
Quadrant III: Urgent, Not Important	Quadrant IV: Not Urgent, Not Important

Above the horizontal line are important things to do, and to the left of the vertical line are urgent tasks. What are some examples for each quadrant?

Quadrant I represents both urgent and important demands on your time: a broken water line, a fallen tree blocking the road, a greenhouse thermal alarm, a baby crying, a barn on fire. Items in Quadrant I are often crises and rise to the top of your to-do list—understandably so.

Quadrant III contents carry urgency but lack importance: certain salespeople knocking at your door, the phone ringing during dinner, and some meetings. Don't confuse urgency with importance.

Quadrant IV items not only are unimportant but fail to carry a sense of immediate action. For me, this can be the quadrant of procrastination: watching TV, reading discount tool catalogs, or surfing YouTube on the computer.

What about Quadrant II? Why did I leave this area for last? This quadrant represents things that are above the horizontal of importance but aren't demanding our immediate action. Because of the lack of urgency, these important items are often relegated to the bottom of the to-do list, including things like long-range planning, business analysis, exercise, creating a great meal, or maintaining relationships. Spending our time on tasks that are important versus unimportant seems like a no-brainer. Yet we are hardwired to respond to urgent matters, whether important or not. The key is to spend your time above the horizontal line of important tasks, and make sure time is devoted to the easily neglected Quadrant II. An easy way to do this is by putting the items on your to-do list and blocking out time for them on your calendar.

Old Decision, New Decision

I recently read a book called *Chasing Daylight* by Eugene O'Kelly. The author was a very successful businessman—CEO and chairman of a Fortune 500 company—fifty-three years old, and at the top of his game. He traveled the world, felt financially secure, and enjoyed time with his wife, daughter, and friends.

One day his wife noticed a funny look on his face, a sort of drooping, and she encouraged him to visit the doctor. After some hesitation, the author saw his physician, who was concerned and asked that he get checked by some specialists the next morning. More hesitation, but O'Kelly consented.

The specialists ran numerous tests and diagnosed a brain tumor that was advanced and inoperable, and they predicted the author had about three months to live. O'Kelly's world had come crashing down in the course of only twenty-four hours.

While the tumor was irreversible, the doctors noted that given the author's particular condition, he would be able to function relatively normally until near the end of his life. O'Kelly quickly stepped down as CEO, relinquished many duties, and focused his energies on his remaining time on this earth. Writing his book was part of his change of focus.

Tragic as his situation was, O'Kelly made the best of it and wrote of his discoveries during his new journey. Common day-to-day concerns paled next to bigger questions, like "What is truly meaningful in life?" With a short and defined life span, he uncovered many gems for those of us who are still living.

One nugget that is relevant to farming, and business in general, is his take on decision making. Ever regretted a decision you made? We all have, but regret no longer. O'Kelly pointed out that the only decisions you should worry about are the ones you haven't made yet. Decisions made in the past were made with all the information you had at *that* time. Any revelations since then are not pertinent. You can't undo the original decision. You only have the power to make a *new* decision.

If you attend a farming conference and find yourself in a lecture not to your liking, you could sit it out and regret your choice of topic. But instead of regretting the decision, make a new decision, and get up and leave. Similarly, I could wallow in regret that I could have spent more time with my young kids while I was starting up my farm. Meeting all the needs of family and work was not easy for me. But with the given situation, I did what I could at the time. I cannot undo the past. Instead of lamenting about what I *should* have done, I decided to spend more time with my family starting *right now*. And I do, and am very happy for it.

Employee Management

The purpose of hiring employees is to help the farmer tackle the farm workload. Quite often, farmers think that each added employee-hour should decrease the farmer's workload by equal amounts. Not so. Even if the employee is as productive in the field as the farmer, time is needed to organize the workday, convey directions, attend meetings, deal with interpersonal relationships, and file mandatory paperwork. Compost happens, but not employee management, at least not without some extra effort. Time and skill are required, and the more of each, the better the management.

I spend lots of time, usually in the morning before the crew arrives, looking at which tasks need to be accomplished that day and in what order. I review what supplies will be needed and create contingency plans in case something goes awry. I also set target times for each job's completion, so I can realistically expect what will get done. Employees like knowing what the day holds for them, whether you work side by side with them or not. Both parties benefit from good preparation, as well as from knowing rough expectations: Weeding one bed should take four hours, trellising one row of tomatoes half an hour, and picking sixty bunches of beets about one hour. If a lot more time is required for a task than was estimated, employees should take note and move on to the next job on the list, if appropriate. I also list contingency plans: If the root digger breaks, move to weeding carrots. Both farmer and employee thus have the list in hand for the whole day's workload, ready to go each morning. I find that old envelopes work great for writing lists. I have plenty of them; they're heavy-duty and fold nicely into a pocket.

Weekly or monthly farm meetings define

work projects and are always welcomed by the crew because they provide a bigger picture of the farm operation. Additionally, the meetings offer a feedback loop to address concerns of the farmer and employee, whether these relate to production or personnel. Regular staff meetings are vital (remember Quadrant II) but can sometimes be put off because of more pressing demands (Quadrant I). Structure staff meetings so that they become a part of your regular work routine.

My formula for maintaining a truly productive and happy employee involves fostering a healthy relationship, one based on mutual respect. I keep away from the traditional labor–management dichotomy as much as possible. *Workers should be viewed not as a liability but rather as an asset.* All too frequently, farmers look at their labor costs as a line item in their budget that needs to be reduced, mainly because labor typically represents a very large percentage of farm expenses. While labor-saving devices are sometimes appropriate for the sake of efficiency, employee treatment should not suffer to bring down labor costs. If you need employees, treat them well. They are fundamental to your farm's overall success. Of all the expenses a farm incurs, wouldn't you prefer to direct the farm's money into the local households of workers instead of the coffers of distant corporations?

There is a huge mixed message we send—whether we are farmers, citizens, or policy makers—in saying that we want to bring back people to the farm, then trying to reduce farm labor to as little as possible. Try looking at it from the completely opposite vantage point: Hire as many people as you can. The community comes to the farm, interacts with it in a meaningful way, and takes part of it back home with them. Alienation between farms and consumers decreases, creating a positive synergistic effect on the farm's vitality. Good farm employment benefits everyone.

I try to make the farm work as enjoyable as possible. I stress that no matter how small or insignificant a task may seem, it is nonetheless *crucial to the overall success of the farm.* I try to break up big blocks of work to avoid burnout and to empower employees. I mix drudgery with more fun jobs. Employees often work together, and in larger numbers, when tackling a big task. Almost no one wants to weed an acre of carrots solo. I hire in extra hands temporarily to keep from falling behind, which is always a boost to the morale of the regular farm crew.

The importance of employee communication and defined relationships cannot be overstated. Written job descriptions spell out responsibilities for each employee and field hand. Defined farm policies clearly address issues like being on time, pay periods, benefits (produce for employees, tool use), workers' compensation, and taxes. When issues do arise, it's easy to let things go unsaid. But unresolved problems inevitably resurface, usually as unproductive sarcastic remarks and blurts of anger. When an issue arises, use *I* statements to express yourself. For example, if an employee is late for work, try saying, "Work starts at 8 AM. I would appreciate you being on time for work. Otherwise I have to repeat instructions for the day, and that is not efficient." Confronting others isn't easy for most of us. Luckily, once the topic is broached, oftentimes it needs no repeating. State your needs and expectations clearly and offer nonoffensive criticism instead of pointed comments that induce shame. Think *negotiation,* and—very important—*listen.*

On the facing page is a job description for any newly hired temporary workers on my farm.

As an employer or crew leader, many times you will need to train *hard workers.* Humans aren't born with a work ethic, but most can change old slow work habits. Slow workers can drag down a whole crew. More productive crew members wonder, "Why should I work so hard when so-and-so is only working at half speed?" I stress to everyone that farming is production work; *the farm gets paid by the piece, not an hourly wage.* In order for the farm to succeed, a certain level of production is necessary, and expected. Cite a target rate that is commonplace for the task: *Hoeing one bed should take forty-five minutes,* for instance. If someone is falling far short of the target rates, work alongside the employee for a while and offer suggestions to picking up the pace. Once new employees grasp the concept that farming is production work, more attention is paid to keeping up.

Hiring extra workers on a temporary basis can make the difference in keeping up with a mounting workload and falling hopelessly behind. I give a standard talk to new temporary employees, often on the phone before their first day of work:

- Work starts at 8 AM. Please arrive a minute or two early so I can show you around the farm (bathrooms, employee space). The farm truck leaves for the fields at eight o'clock sharp.
- Bring a water bottle, lunch, and appropriate work clothes (and a hat and gloves depending on the weather), ones that you don't mind getting dirty.
- We will break from noon to 1 PM for lunch behind the house. Work ends at 5 PM.
- Payday is Friday or the last scheduled workday of the week.
- Pay is $10/hour. Timesheets are in the employee cabinet. Keep track of your own hours.

I hand new employees a Cate Farm Crew job description when they arrive on the farm and make sure any questions are answered. If I am not going to the field with the crew, I let everyone know who is the designated crew leader.

Regular Employees

Regular employees commit to work on a consistent basis. Once a certain threshold of pay is reached, tax laws and workers' compensation become mandatory. Circular A from the Internal Revenue Service defines the thresholds

JOB DESCRIPTION: CATE FARM CREW

Welcome to Cate Farm! We appreciate your hard work and want to have clear policies so that both employer and employee understand each other and mutually benefit.

Work at Cate Farm is diverse, ranging from greenhouse seedling production, greenhouse tomato culture, field crop planting, weeding, and harvesting to building and repairing farm infrastructure. Employees are often called upon to do different and sometimes tedious jobs in all kinds of weather. Be prepared for hard physical labor. Job training will be done by Richard, Sally, or Pete. It is important to remember that no matter how simple, tedious, or insignificant a task may appear, everything that you do is important to the overall success of the farm.

Farming is production work. The farm earns money by what is actually produced and sold. Since production per hour or day is very important, employees are expected to work quickly and efficiently, to keep up with the pace set by the employers or to work quickly on their own.

Conversation during work is a benefit of farm work, but please be aware of keeping your hands moving while you talk. Efficient production is critical to the success of the farm.

Paid time is work time. On a farm there is rarely nothing to do. If a task is completed before another task is assigned, make use of your time by doing a job on the pack house bulletin board such as cleaning up, weeding ends of rows or greenhouses, or making tomato boxes.

Employees may be asked to work around machinery and should exercise caution when doing so. No one will drive a tractor unless trained by Richard. Farm vehicles will be driven by regular employees only.

At Cate Farm we ask that you:

- Be punctual, and be ready to work at the designated time.
- Bring a water bottle, lunch, and appropriate clothing for the weather and task.
- Keep track of your hours on a Cate Farm timesheet.
- Telephone if you are unable to make it to work.

Paydays are the 15th and last day of each month for regular employees, and the last day of the week for temporary employees.

There will be no alcohol or drug use while working on the farm. On the job, it is expected that all employees will behave appropriately and cooperate with other employees and people on the farm. Employees are expected to keep work areas and fields clean by picking up trash and tools associated with the work.

If you have an issue or problem on the job, please bring it to the attention of Richard or Sally. We want to make things work for all concerned.

We are cooperatively minded and strive for a harmonious workplace.

Thanks for your help.

and explains tax requirements in detail. Workers' compensation is an insurance, not an IRS program, that employers purchase to cover on-the-job employee injuries. Your local insurance agent will be able to help you out. But to give you an overall idea of what's involved, I'll briefly describe the requirements if you decide to hire regular employees.

The IRS requires any eligible employer to:

- Obtain an Employer Identification Number (EIN) from the IRS.
- File forms W-4 (withholding allowance) and I-9 (employment eligibility) for each employee.
- Keep track of employee wages.
- Deduct employee taxes from paychecks on behalf of the employee (withholding, Social Security, Medicare) and deposit them into an authorized financial institution, like your bank, for the IRS.
- Pay (from your farm checkbook) to the IRS a matching amount of each employee's Social Security and Medicare contribution.
- File Form 943 annually, documenting your payments to the IRS.
- File annually each employee's W-2 form, and the total for all employees, Form W-3.

Doing all this paperwork takes extra time, and your financial contribution to each employee's Social Security and Medicare currently adds up to 7.51 percent of payroll. I'm not aware of any other "volunteer" work aside from that of the IRS that requires you to perform certain tasks, or risk going to jail or losing your farm. Business owners are not compensated for their added duties, and the IRS has teeth. I can joke about it now, but at first I was a bit overwhelmed. I understand the need for employee taxes and corresponding laws to hold employers accountable. Businesses are an appropriate vehicle for making it all happen. Payroll services, from either independent companies or computer programs, can save time and possible aggravation, but usually they come with an added price tag. I included a Payroll Calculator on the CD accompanying this book so you can simplify and speed up payroll processing. Read through Circular A from the IRS or talk to your accountant to be well informed on employer tax responsibilities.

Workers' compensation insurance is separate from the IRS and covers your employees for work-related injuries. Policies are available through many insurance companies. Costs per employee vary depending on the type of work performed and your track record with reported injuries. Figure on spending about 8 percent of payroll. My home state of Vermont holds employers accountable for worker injuries even if no insurance is purchased. Annual premiums are calculated from your annual payroll and documented to the insurance carrier with IRS Form 943. Sometimes an auditor from the insurance company will visit to verify your payroll records.

Many beginning farmers will find these employer mandates a challenge. The steep learning curve combined with the added costs may convince farmers to fly under the regulatory radar and pay employees a simple hourly wage. While more than a few employers take this route, consequences exist, as some celebrities who hired off-the-books domestic workers have found out. I don't advocate this as a sound business practice, but I can relate to the sometimes overwhelming complexity of starting up a farm. Table 6-1 lists all these mandatory employer duties to give you an idea of all that is required.

Workers' compensation policies are often written by the insurance company annually. Payments are usually annual or quarterly. Audits, if any, most likely occur only once a year.

SEP IRA

Individual retirement accounts (IRAs) are a tax-advantaged way of building a nest egg for retirement, and they come in different forms. *SEP IRA* stands for simplified employee pension individual retirement account. This type of company-sponsored retirement plan for businesses was designed to be simple for small businesses to administer. You can contribute to a SEP IRA plan in addition to your own personal Roth IRA or traditional IRA. A SEP IRA plan serves owners and employees alike, and even a business with no employees benefits from this kind of IRA. SEP IRAs are a profit-sharing plan in which farmers and eligible employees take up to 25 percent of their pretax earnings and deposit the money to grow without taxation. The SEP IRA annual percentage rate is variable, set each year by the farmer. If it was a good year, the maximum rate of 25 percent can be applied. If not, a lower or 0 percent allocation can be used. Because of the tax benefit, I aim for the maximum 25 percent each year. Whatever percentage is chosen, the farmer and all eligible employees share the same rate. Here's how it works.

The farmer fills out a simple IRS form (5305-SEP) for each eligible employee. The form is kept by the employee. No paperwork

TABLE 6-1: Calendar of Employer Duties

(Note: This is intended only as a guide; check for the latest IRS information.)

January 1–February 15	Ask for a new Form W-4 from each employee.
By January 31	File Form 943 to the IRS for the previous year's employees.
By January 31	File Form W-2 for each of the previous year's employees and furnish them a copy.
By February 28	File Forms W-2 and W-3 (the total of all W-2s) to the IRS. Check for your state's filing dates.
Each month (depends; check IRS Circular A)	Deposit employee and employer shares of Social Security, Medicare, and federal income taxes withheld to the IRS via an authorized financial institution (usually your bank).

is sent to the IRS. To be eligible, an employee must have worked at least a small amount per year in the last one to five years. Parameters are set by the farmer. See the IRS form and Publication 560, *Retirement Plans for Small Business,* for more details. The farmer sets the percentage rate for the SEP IRA at year's end. That percentage rate is multiplied by the farm's annual net profit. The resulting amount is written as a check from the farm account to a SEP IRA custodial account in the name of the farmer. The same percentage rate is also applied to the annual gross wages of each eligible employee, with a farm check written to a custodial account in the employee's name. The SEP IRA accounts can be opened with numerous financial institutions—CDs at banks, mutual funds, or insurance annuities, for instance. Whether written to an account for the farmer or the eligible employee, the SEP IRA checks are qualified business expenses. The farm's net profit is reduced, resulting in tax savings. Moreover, SEP IRA payments to the farmer and employees are not subject to employee taxes or workers' compensation, a savings of about 15 percent.

Enrolling in a SEP IRA is easy and advantageous to the farmer. Without employees, the farmer reaps the benefits as the sole recipient. With eligible employees, the farm's added cost paid to their SEP IRA accounts is part of their employment package, a very real part of the employees' compensation. If employees' eligibility kicks in after their second or third year of employment, the added SEP IRA benefit is an incentive to keep them coming back to work each year.

Farming with Your Spouse

Anyone farming with their spouse will understand why this subject needs to be talked about. I conclude this chapter of effective management with this topic because dual management of a farm adds a layer of complexity to everything written above.

Maintaining a healthy relationship with your "significant other" requires time, communication, and openness. Relationships take work to thrive. Many nonfarm couples see each other only before and after the workday. Couples managing a farm are together 24/7. This requires even more communication, openness, and time. The need for negotiation is ever-present in order to craft decisions jointly. Power struggles develop. All this can place stress on a relationship.

My wife, Sally, and I realized that most conflicts arose when our job boundaries were fuzzy or overlapping. Each of us has our own niche on the farm and performs tasks that we prefer to do or are better suited at. Although we work together all day on the farm, each of us has our own domain. For instance, Sally is in charge of all the greenhouse production, whereas I am responsible for field crops. But conflicts arise when employees are urgently needed for both greenhouse and field work. Who recruits new employees? How is employee time spent? Sally and I have to work out the answers. We spend a lot of time communicating our needs

SOCIAL SECURITY BENEFITS

If you and your married spouse operate your farm business as an IRS-defined "sole proprietorship," only one person will accrue Social Security benefits for the net farm income. After many years that means one spouse will have all the Social Security wages in her or his name, while the other spouse has none. That may not sound fair, but that is the way it currently works. In striving for more equality, some couples alternate the benefit each year or pay one of the spouses as an employee. But are those the best tactics? If all the benefits are in one spouse's name and then he or she dies, often the remaining spouse will receive more, not less, in Social Security payments than if each spouse had an equal contribution. Talk to your accountant to find out the best scenario for you.

to arrive at agreeable solutions. Making any decision takes more time when co-managing than when doing it on your own, whether for labor needs or farm purchases. We have regularly scheduled meetings for a weekly overview, to discuss the ongoing list of tasks to do and who will do them. Daily morning check-in meetings, even if brief, help us center ourselves. Conflict can be scary, but don't let that fear keep you from talking a subject out to its resolved ending. Put the subject on hold if you need some more time to see a clear solution. Lean on the safety of your commitment as partners to honestly express yourself. In addition, the willingness and ability to compromise is essential to negotiation and joint decision making.

Another related aspect of working with your spouse is the boundary between farm and home. If your home is on the farm, it is easy for the two to meld into one. Some farmers find that a non-issue, as I did when first starting out, but over the years I've come to appreciate keeping some areas in the house separate from the farm. The "home" is a sanctuary, a place of relative privacy and relaxation. Bedrooms typify this quality; not everyone is invited inside. The farmhouse may be your home, but it often includes an office, a bathroom, and a kitchen for crew meals. My wife and I enjoy aspects of the farm in our home and choose to relegate other parts outside. To maintain a sense of home, we provide a different bathroom and area for employees and ask that these spaces be kept clean. Lunches are on our porch, or in our kitchen if it's cold or rainy. Employees are informed of our desire to keep our home and farm separate, but not in an unwelcoming way. Each farm couple has different feelings about home and farm boundaries; I just want to encourage you to talk about it.

– 7 –

Office Paper Flows and Leaky Finances

Office management is another humdinger of an exciting chapter subject, right up there with record keeping or planning for your own death. But as any small business that has successfully cleaved its niche into the marketplace has discovered, proper office procedures are as important to overall success as marketing and efficient production. Frustrating piles of bills and invoices can become a deterrent to tackling office work and, worse, can easily erode your bottom line—sometimes without you even knowing it. With the advent of personal computers in the 1990s, I hired a bookkeeper to input all my paper invoices onto a software program. She found $800 in uncollected payments and math errors. Eight hundred dollars! Granted, that represents only one-half of 1 percent on $150,000 gross sales. Still, if I dropped $800 on the ground, you can bet that I'd take time to pick it up—that's no couch change! This chapter addresses the procedures needed to run an efficient and tight office that saves you money.

One study of businesses revealed that 75 percent of business errors were the result not of human error, which is so often the scapegoat, but rather *procedures*. I always thought most mistakes were made by humans—so how could procedures be the culprit?

Take the airline industry, for example. A single plane crash ruins an airline's reputation for safety, can cost hundreds of thousands of dollars, and puts into question the company's ability to survive. Plane crashes are bad for business. Yet thousands of flights take off and land every day without incident, despite the multitude of employees that are prone to making a mistake. How so? Airlines recognize that human error exists, so procedures are set up to prevent them from happening. Double and triple checks prevent one person's mistake from wreaking havoc. Written logs document checkpoints along the way. A possible oversight by one person is caught by one or more safeguards that are built into the system.

The same is true with the hospital business. One story of a patient's wrong foot being amputated not only is embarrassing, but severs business as well. I accompanied my father to the hospital for a pacemaker replacement and no less than four different hospital employees (the check-in desk attendant, the nurse, the anesthesiologist, and the surgeon) asked him directly what procedure was to be performed—and listened to his answer. A little repetitive, but it left no doubt in anyone's mind. Another example that hospitals use to eliminate human error is called the "Five Rights of Medication Administration": Right Patient, Right Route, Right Dose, Right Time, Right Medication.

Back on the farm, too, everyday procedures prevent human error. Writing notes for phone messages makes certain that the intended person receives the communication. Thermal alarms warn greenhouse operators of heating or cooling malfunctions. Timers on lights prevent unnecessary electricity costs. Tractor safety training and handouts clearly spell out the dangers of a spinning PTO shaft. Duplicate paper invoices make sure payment is received for product sold.

Paper invoices are a standard practice for good reason—without them sales revenue is easily lost. Remember my $800? Say I'm selling a washed and graded 25-pound bag of carrots from the farm to a community member for $25 and that person forgets her checkbook. She promises to mail a check, and I trustingly agree to it. All is fine so long as one or both parties remembers, but in our sometimes hectic lives, details are forgotten. Days pass, and no check, and the transaction slowly fades from each of our minds. That nonpayment is a direct loss to my bottom line—*it is as if I took $25 out of my wallet and tossed it to the wind.* Gone. All my expenses for growing and packing out the carrots, however, are very real and not gone with the same wind. This loss of funds falls where it hits hardest—from your net profit.

Along the same lines, I often lend out reference books, cultivation tools, and other items to neighboring farmers, in the spirit of cooperation and helpfulness. I trust these neighbors will use the borrowed items and return them, but the onus is on me to remember to get the goods back. A lost book or tool can cost me $50 or more to replace and is no different than the lost sales from a bag of carrots—it comes directly from my wallet.

Twenty years ago, I loaned a new Havahart trap for skunks to someone, never got it back, and have yet to figure out who it was. Cha-ching, that's another $50. Since loaned items aren't an invoice type of transaction, I now simply keep a sheet of paper on my office wall to record when any book, tool, or CD leaves the farm, who took it, and when. A paper trail, whether it's a piece of paper on the wall, a logbook, or a copy of an invoice, guarantees against loss from human forgetfulness.

Two distinct paper trails flow in the farm business. One trail, purchases, tracks all items or services coming *into* the farm (such as seeds, potting soil, and fuel, each with a corresponding bill to be paid); the other, sales, monitors all products *leaving* the farm (carrots, beets, or parsley, each with an accompanying invoice, entry into a CSA logbook, or farmers' market inventory sheet). These two paper trails will be dealt with separately, with the goal to provide you a step-by-step procedure to guarantee no lost sales revenue and maximum tax benefit for all your hard work. The first to tackle are purchases: items or services bought for farm use.

Purchases

I'm a firm believer in bundling tasks for efficiency, whether that means picking all the cilantro for all the account orders first and then picking all the parsley; folding all the boxes needed to fill orders first and then packing them; or filling all the needed seed flats before seeding them. Office tasks are no different. I work in blocks as much as possible to maximize efficiency and reduce human error. Set up a regular time once a week, such as Wednesday 9 AM to noon, to devote to desk work. Biweekly schedules work fine as well, but weekly habits are, well, more habitual. If you don't have the time, find someone who does.

Three types of purchases are common: ordering items over the phone or Internet and taking delivery of them (like seed); driving to the store and paying or charging the item (like fuel); or services and items that come to your door (like grid electricity). In all cases, the business purveying the goods or service supplies paperwork: a bill to be paid, a charge slip, or a receipt for paid transactions.

The simplest office setup for purchases uses two folders: one for paid bills and one for unpaid. If I pay a bill immediately, as I sometimes do at a store, I record the date and check number on the bill. This paid bill is placed into a pocket folder labeled PAID BILLS in my office desk drawer. Future paid bills are placed behind previously paid bills, so all the bills are in chronological order of payment. This simplifies finding any paid bill in the future, if needed. All my paid bills for Cate Farm for the year fit into one pocket folder (albeit tightly).

Unpaid bills come in the mail, with deliveries, or from businesses that give you credit. Any unpaid bills go to my office and are placed in a holding area in plain sight, like an upright paper holder. I process unpaid bills on a weekly schedule. No need to pay them one by one every time you open the mail. Bundle the task of paying bills and process them all at once.

Gather all your unpaid bills. Take the first bill, make sure it is accurate, and write a check for it. Record on the bill, "paid check # _____," with the date, and circle the amount of the bill to denote what you paid. If you are not paying the full amount of the bill, record the amount that you did pay. Now place your check and the bill's payment stub in an envelope, seal it,

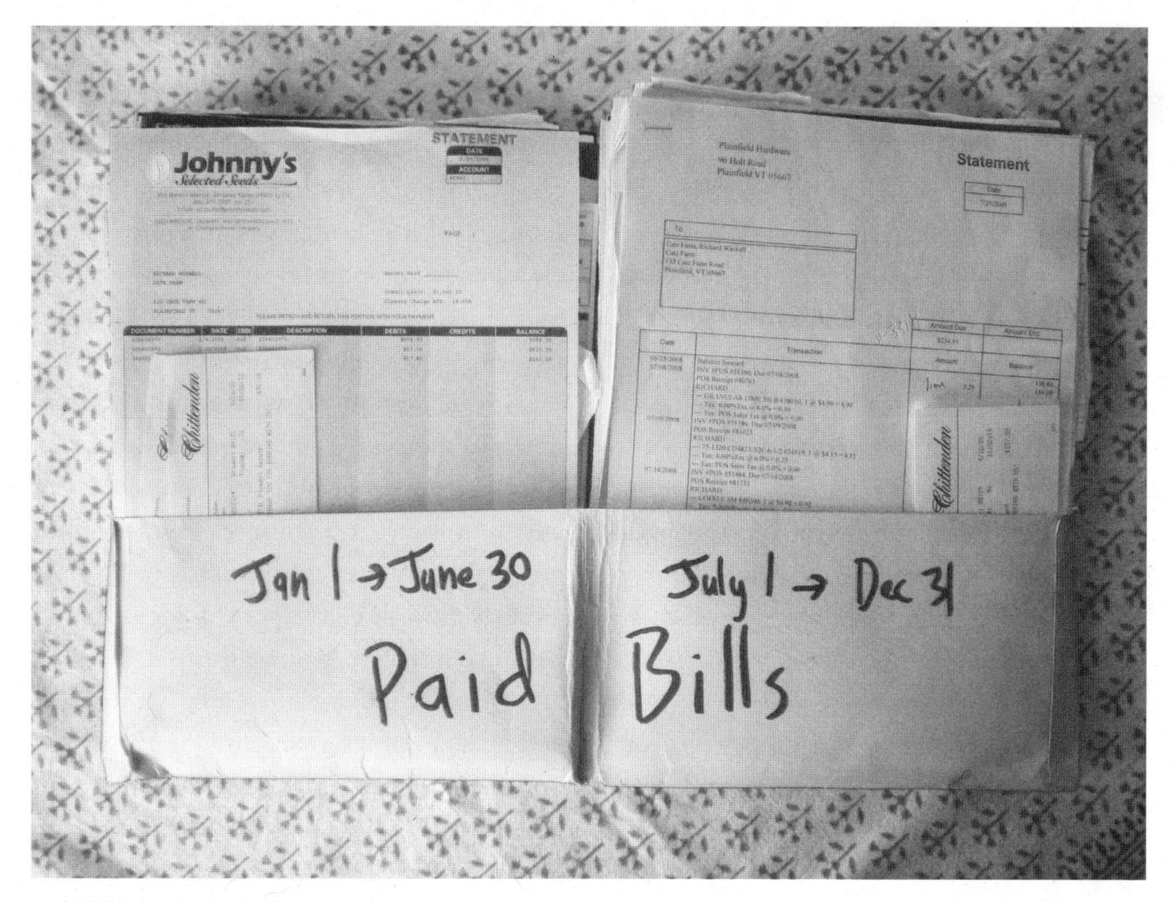

Paid bills in chronological order.

and set it aside for later postage when all the bills are paid (again, bundling tasks). The paid bill now is placed behind the other paid bills in the pocket folder in your desk drawer. Finish paying all your bills and then place postage and a return address on each envelope. If some bills are not due immediately, or if cash flow is tight and some bills can take a backseat, return them to the unpaid bills holder for future check writing.

A few variations on this simple paid bills/ unpaid bills setup warrant discussion. Some are commonplace, like using personal checks or cash to pay for farm expenses. Others will increase office efficiency, such as paying by credit card or setting up monthly statements.

Using Cash

If I'm delivering produce in town and running on empty, but I forgot my checkbook, I may pay cash for some fuel to get home. Because the delivery fuel is a bona fide farm cost, I want to make sure I expense it through the farm account to take advantage of the substantial tax benefit. Chapter 8 explores this topic in detail, but, in short, if I fail to record a farm expense, I personally lose one-third of the purchase price.

Say what? If I spend $45 on delivery fuel and forget to record it as a farm cost for tax purposes, I don't get to offset my farm income with that $45. My year-end profit would be $45 higher, and I'm going to have to pay taxes on that extra $45. Taxes for the self-employed cut a full third out of year-end profit (usually 15 percent federal income tax, 15.3 percent Social Security and Medicare, and roughly 4 percent state income tax). So that's an extra $15 I would have to *pay* in taxes just because I didn't record my $45 in fuel as a farm expense. *Every farm expense should be used to offset your farm income and profit to save overpaying taxes.*

This example shows that I'm actually $15 poorer by not recording the $45 of fuel (paid by cash, or any other means) as a farm expense. Fifteen dollars may not seem like a lot, but over the course of a year, all missed expenses add up—and come right out of your wallet.

The lesson: Record all your farm expenses, cash or not.

What do I do when I pay cash for a farm item? I write on the slip "paid cash" and put it *into the unpaid bills area* in my office. I pay that bill from farm funds as I would any other unpaid bill, the only difference being that I make the farm check out to myself—reimbursing my personal funds. I mark on the bill "reimbursement for ____, paid check # ___," add the date, and circle the dollar amount.

Set Up Accounts Billed Monthly

Another variation on the basic paid bills/unpaid bills setup is one that increases efficiency and reduces errors: setting up expense accounts with monthly statements for all the businesses you routinely purchase from. Hardware stores, auto parts outlets, feed suppliers, seed companies, and gas stations are all likely candidates. Instead of writing a check every time you purchase gasoline, or some nuts and bolts, or a packet of seed, let the business you buy from keep a tab going for you, and then pay by monthly statement. You'll never write more than twelve checks per year for each business, and you'll be writing them at the calm of your office desk—not rushing off to finish deliveries or bolt up your irrigation pump.

Before I set up an account with the hardware store, I'd be writing dozens of checks for smaller amounts, plus paying cash out of personal funds. Now I get what I need at the store and sign the bill (keeping a copy), and I'm off. When I get home, I put the unpaid bill in a separate paper holder labeled ACCOUNT SLIPS near the UNPAID BILLS holder. When I receive a monthly statement, I match up the corresponding account slips, staple them to the statement, and place it in UNPAID BILLS or pay it immediately. By setting up expense accounts and paying by monthly statements, you will write far fewer checks and reduce errors by being more focused.

Credit Card Purchases

The last variation, paying by credit cards, can also increase efficiency and reduce errors—but only if you don't have an issue with overspending. Credit card interest rates are often exorbitant. Credit cards must be paid in full every month; if not, the interest payments will be a big leak in your profitability pail (more on that in chapter 8).

Credit cards are accepted nearly everywhere and can be used for personal and farm expenses alike. When purchasing by credit card, keep the slip and bring it back to the office to put in a separate holding area labeled CREDIT CARD SLIPS, near UNPAID BILLS and ACCOUNT SLIPS. When the monthly credit card statement arrives, staple all the corresponding slips to the statement and place in UNPAID BILLS or pay immediately, the same process used for monthly accounts described above. I use the same credit card for personal as well as farm business expenses. I write two checks each month to the credit card company, one from my personal checking account and one from the farm checking account.

Designate a Mother Checkbook

A household that also runs a small business might keep one, two, or more bank accounts: a personal checkbook, a farm checkbook, a savings account, a money market account, and a certificate of deposit (CD). If no separate farm checking account exists, I'd recommend opening one at some point in the future. A personal checkbook can work to manage finances for farm transactions, but there are advantages to having separate accounts.

The farm checkbook tracks most of the farm transactions—designate this account as the "mother checkbook." Record *all* money coming in and going out from the farm in this checkbook's register, so there is only *one* place to track all the money flow.

Doesn't this happen automatically? Not necessarily. If funds in the farm checkbook are temporarily low (never happens!), you may be inclined to pay off a large loan bill or credit card statement with funds from your personal checkbook or savings account that are more flush. That's okay to do, but it can be confusing when sorting out farm finances at year's end. A more foolproof tracking method would be to write a personal check to the farm account and then pay the farm expense with the farm checkbook. That way there's no need to scour the office to determine if something was paid or not, or if money was paid or received—it's all right there in the farm checkbook register.

Maybe it's just me, but for a long time I thought that only two lines could be used to describe each transaction in the smaller checkbook registers (one line shaded, one not). So I'd fill it in as shown in table 7-1.

But one day a flash of inspiration hit me. Unlike my checks, which I have to purchase, my bank *gives* me the registers for free! I can use more than two measly lines—maybe writing details on four, six, or even ten lines if need be! I started busting out of the two-lines-only mentality and started recording all the information that was relevant, and where it was most useful. Now I itemize each different type

TABLE 7-1

Date	Check #	Description	Debit Amount	Credit Amount	Balance
7/2	5467	Feed store	60.38		
		June bill			
7/3	5468	Visa statement	832.62		

TABLE 7-2

Date	Check #	Description	Debit Amount	Credit Amount	Balance
7/2	5467	Feed store: $20 fence wire			
		$30 oats for cover crop			
		$10.38 canine fodder/personal	60.38		
7/3	5468	Visa: $45 gas for '97 truck			
		$238.60 HM Seeds			
		$280.35 van leaf springs			
		$80 NE conference			
		$188.67 plane to Mom's/draw	832.62		

of farm (and nonfarm) transaction right there in the checkbook register, even if it takes a whole page. In this way, I can see how the feed store bill and credit card statement truly sugar off (see table 7-2).

When I want to total up an expense category, every transaction is in one place. Year-end bookkeeping is simple and accurate. There's no need to leaf through each paid bill one by one to categorize expenses. I also don't worry about missing an expense, which as I earlier showed can rob you of one-third of the expense transaction.

Breaking apart expense types in your checkbook register does take more time than just writing a check for the dollar amount, but it is time saved in the long run. Again, if you designate a specified time each week to work at your desk for bill paying, you will be at the calm of your desk, without distractions, and less likely to make errors.

Setting Up Expense Categories

We track expenses to monitor where all our hard-earned money goes. The amount of money coming into our checkbook is limited, so we have to make choices to spend money on some things and not others. Knowing the overall amount for each type of expense helps determine future spending choices.

What kind of expenses should be tracked? Ones that are meaningful to you for managing your business. Every farm is unique and each manager an individual. There is not *one* universal list of farm expenses, but certain forces shape the expense listings, like the tax code. Creating expense accounts to effectively manage your business should also dovetail with tax filing requirements. Like it or not, taxes are a part of business, so why not make it as easy as possible to figure your deductible expenses? Below is a list of possible expense accounts:

- Advertising
- Donations
- Fertilizer
- Fuel and oil
- Greenhouse supplies
- Insurance: farm share
- Interest expense: farm share
- Internet, website
- Livestock expenses
- Miscellaneous
- Office
- Payroll
- Professional services
- Rent paid
- Repairs and maintenance
- Seeds, plants purchased
- Supplies
- Taxes
- Trucking, freight
- Utilities

Any of these can be changed and broken down into subaccounts for more accurate monitoring. For instance, you may want to differentiate between types of fuel used:

Fuel and Oil
- Off-road diesel
- Gasoline for tractor
- Propane for greenhouse
- 1997 truck gas
- 1995 van gas
- Other

When expense accounts are coupled with income accounts, a title page called a Chart of Accounts is created. Below is an example with expanded expense accounts:

CHART OF ACCOUNTS

Income

Sales:
- Wholesale
- CSA
- Farmers' market

Expenses

Advertising

Donations

Fertilizer

Fuel and oil
- Off-road diesel
- Gasoline for tractor
- Propane for greenhouse
- 1997 truck gas
- 1995 van gas
- Other

Greenhouse supplies
- Pots and trays
- Potting mix
- Building supplies
- Other

Insurance: farm share
- Vehicle
- Farm policy
- Workers' comp
- Health

Interest expense: farm share
Internet, website
Livestock expenses
- Bedding
- Feed
- Other

Miscellaneous
Office
Payroll
Professional services
Rent paid
Repairs and maintenance
- 1997 truck
- 1995 van
- Pack house
- John Deere
- Ford tractor
- Cub
- Shop supplies
- Other

Seeds, plants purchased
Supplies
- Labels, boxes, bands, etc.
- Other

Taxes
- Property tax
- Sales tax

Trucking, freight
Utilities
- Electricity
- Landfill
- Telephone

Expense accounts and subaccounts keep track of costs in as much (or little) detail as you want. Each category can be used for tax purposes, by placing your expenses into the appropriate line of the IRS Schedule F. Some expenses match up directly with Schedule F,

TABLE 7-3: Greenhouse Supplies

Date	Explanation	Pots and trays	Potting mix	Building supplies	Other
1/15/09	10 yards potting mix		1,050.00		
1/30/09	12 cases 1020 flats	648.40			
2/7/09	2 hoses, wand, shutoffs				97.96
2/8/09	1 x 4s for benches			35.60	
2/15/09	4 cases 806 packs	256.78			
2/16/09	Thermalarm for greenhouse #2				53.20
	Subtotals	905.18	1,050.00	35.60	151.16
	Total expenses: $2,141.94				

and some may need to be combined. Table 7-3 shows an example of the expense account (also called a ledger page) for greenhouse supplies.

You can transfer your expenses from your mother checkbook to paper ledger pages or enter them into a computer bookkeeping program. All your expenses are tracked the way that is most useful to you. Tax preparation is made simpler, and the accounts can be used to generate lots of useful information for analyzing your business, such as a Profit and Loss Statement, a Balance Sheet, and Cash Flow Projection (covered in chapter 10).

Summary of Purchases and Bill Paying

- The basic setup contains unpaid bills and paid bills.
- Variations include:
    - Cash reimbursements—unpaid bills, payable to yourself.
    - Setting up accounts and paying by monthly statement—account slips.
    - Credit card purchases—credit card slips.
- Reap tax benefits by expensing cash and personal check purchases.
- Bundle tasks.
- Designate a mother checkbook.
- Use lots of register lines to break down expense types.
- Write fewer checks, in the calm of your office.
- Set up expense categories and subaccounts.

Computer-Generated Checks

On the one hand, computers can be seen as incredible tools to help process information; on the other, they have the potential to alienate users from other people and the natural world—so much so that the computer might someday end the world as we know it. Computer bookkeeping software, however, is definitely on the "incredible tool" end of this spectrum.

Many programs are available at affordable prices, and they are a boon to the small businessperson. Bookkeeping software can store information, sort it in any number of ways, generate reports, process taxes, reduce errors, and eliminate duplicate tasks. Depending on your computer savvy, the learning curve may be a little steep, but it levels out quickly after your initial efforts.

The process of bill paying by computer is inherently the same as with handwritten checks. I use QuickBooks, the most basic edition. The graphic art on the computer screen depicts checks and invoices as you normally envision them. Checks are written on the computer screen, and then a printer fills out the check on a paper check template. I still need to sign the checks, but the payee's name and address, as well my name and return address, make it a snap to mail in a window-style envelope. Plus, check expenses are instantly put into the appropriate expense account in your Chart of Accounts.

I still keep my old write-by-hand checkbook to bring with me to the store and on other errand runs. It is drawn on the same bank account, but with a different-style check and check numbers. I enter these handwritten checks into the computer (my mother checkbook/farm account), where they are recorded in the register with all the other checks. The monthly bank statement lists both sets of numbered checks that have been cashed, side by side.

Reconciling Checkbooks with the Bank Statement

Fear no longer the dreaded task of balancing your checkbook! I used to put it off, anticipating frustration at every turn. I wondered why I had to be subjected to this and how necessary it was in the first place. The answer to both questions is simple: You, and only you, are responsible for keeping track of your money.

I've never been that good at doing crossword puzzles. I complete all the easier answers first and then usually give up before all the blanks are filled in. Even though I don't finish the puzzle, *I know that a solution exists.* With enough time and help, *I could complete* every blank as intended. Checkbook balancing is just a different crossword puzzle.

"Close enough for government work" is an oft-used saying in many trades, but not by bookkeepers. Checkbooks are balanced to the penny, partly because they can be. Like the crossword puzzle words, every penny will fit into its proper place. And armed with the knowledge that the correct answer indeed exists, balancing your checkbook can become a challenging puzzle instead of a frustration. Here's how:

I use a system of X's and O's—not shorthand for hugs and kisses, but rather canceled and outstanding checks. When a bank statement arrives, I first label the envelope with which account it is and which month and year (FARM FEB '09) for future ease of reference. I open it and start by marking off in my checkbook register (on the computer or in the paper register) the checks that the bank lists on the statement. I put an X next to each canceled check the bank lists. If a check that I wrote (and duly recorded in the register) has not yet been cashed, the bank doesn't know about it, and it won't be on the statement. I place an O next to checks in my register that have not yet been received by the bank. These O's are outstanding checks within the bank statement's time frame (and will eventually have an X marked inside them with the next month's statement when the checks clear). I also place an X or an O next to each deposit I put into the bank for the same time period.

After the last X or O in my register, I draw a *squiggly line* (that's a technical accounting term) to mark the end of that month's bank statement. Just above this line, I enter any bank charges or interest from the bank statement that I previously didn't know about but need to include.

I calculate the balance for the checkbook register as I normally would and enter the balance just above the squiggly line. That amount should equal the bank statement ending balance, with outstanding checks added in (and outstanding deposits subtracted). See table 7-4 for an example.

Tables 7-5 and 7-6 show elements of a typical monthly bank statement.

My register balance is $1,912.23. Adding outstanding check #880 for 256.78, the total is $2,169.01. This matches the bank statement's balance—hooray!

But what if the two amounts are different? Think like the bank statement. *Be* the bank statement!

TABLE 7-4

Date	Check #	Description		Debit	Credit	Balance 3,465.70
2/3/09	876	VT Compost Co.	X	1,050.00		
		10 yards potting mix				2,415.70
2/6/09	877	Griffin Greenhouse Supply	X	648.40		
		12 cases 1020 flats				1,767.30
2/7/09	878	Plainfield Hardware	X	97.96		
		2 hoses, wand, shutoffs				1,669.34
2/8/09	879	P&R lumber	X	35.60		
		1 x 4s for greenhouse benches				1,633.74
2/15/09	880	John Smith	O	256.78		
		4 cases 804 packs				1,376.96
2/15/09	---	Deposit: Co-op sales	X		537.00	
						1,913.96
		Service charge and interest		2.35	.62	1,912.23

TABLE 7-5: Bank Statement, Period Ending 28 February 2009

Beginning balance	3,465.70
Debits	1,831.96 (John Smith hasn't cashed his check yet)
Credits	537.00 (deposit: co-op sales)
Service charges	2.35
Interest	.62
Ending balance	2,169.01

TABLE 7-6: Checks

Date	Check #	Amount
2/7/09	876	1,050.00
2/9/09	877	648.40
2/10/09	878	97.96
2/13/09	879	35.60

The bank knows only what it received and doesn't often make mistakes (but it can). First, check *your* math. Are you starting from a true point of reference? Recalculate the register balance. Recheck the bank's canceled checks and deposits. Retrace your original steps. Still off? What is the difference in amounts? Is it recognizable as a check or deposit amount (or double or half that amount)? Remember the crossword puzzle with its definite solution. If you need to, walk away from the problem for a while, or ask a friend or the bank for help. A solution is out there.

As a shortcut if you are still completely mystified, assume that the bank is correct, use its balance, adjust your register, and start with the next month's statement to balance the register with what the bank says you have. You will soon be a crossword master.

Sales: Money Coming into the Farm

Chapter 3 explained some simple ways of tracking different sources of income. Invoices are used with wholesale accounts, CSA shares are recorded in a notebook, and farmers' market sales are documented with an inventory sheet for each market. All farm products leaving the farm are recorded in one way or another. But what about payment for those products? How do you make certain that you are getting paid for the products leaving the farm? What is the most efficient way of processing the sales paper trail?

Deposits

A place for everything and everything in its place. Proper procedures reduce errors. In chapter 6, I described various holding areas in your office for pieces of paper like checks and invoices. Now it is time to process these papers.

When I receive a check (or cash) upon delivery for an invoice of farm products, I write "paid check #___" or "paid cash" and place the payment and invoice together in a holding area for payments. I use an upright paper holder on my desk in plain view. If a payment is not paid upon delivery, I write "Not Paid" on the invoice and put it in an alphabetical accordion folder of unpaid invoices, also in plain view. Farmers' market sales money and CSA payments go directly to the upright paper holder for payments along with a piece of paper explaining the payment. To repeat, I have my upright paper holder for payments (with a corresponding paid invoice or paper attached) and an accordion folder for unpaid invoices.

When a check does come for an unpaid invoice or invoices (usually in the mail), I place the check in the upright paper holder with the other payments. When I'm ready to make a deposit to the bank (usually once a week), I first retrieve all the checks and cash from my upright paper holder. Next I take a blank piece of paper as a cover sheet for the deposit. This deposit title page is usually a folded 8½ x 11 paper from the recycle bin, which is about the same dimensions as my farm invoices. I write FARM DEPOSIT and the day's date at the top and list all the checks and cash payments below the

Unpaid invoices (right) and payments for deposit (left).

title. I write account names like Hunger Mt. Co-op followed by the amount of the check or checks. Farmers' market and CSA payments each have their own heading. Small accounts or those I sell to infrequently are listed under the heading MISCELLANEOUS SALES, each with an accompanying explanation. See table 7-7.

Now I need to match up each payment with the paper trail I use to track all sales from the farm. This ensures that every product is paid for. Some parts of the paper trail are attached to the payments in the upright folder, like invoices that were paid at the time of delivery or the inventory sheet with sales from the farmers' market. Some payments, such as the checks that came in the mail, still need their paper trail companion. I look into my alphabetical file holder, where all my unpaid invoices are stored temporarily, and match up the payments with the appropriate invoice or invoices.

For instance, when I receive in the mail a check from Hunger Mountain Co-op for $680.50, it is placed in the upright paper holder payment area. When doing a deposit, I take that check (along with all the other payments), look under H in my accordion file, and pull out all of the

TABLE 7-7: Deposit Title Page

Farm Deposit 7/23/09		
Hunger Mt. Co-op	680.50	
	432.00	
	560.65	
Farmers' market cash	970.00	
Checks	126.32	
CSA late payment	225.00	John Doe
Miscellaneous sales	50.00	check for 2 carrots
	15.00	cash for tomatoes
	123.45	town fair, with invoice
	Total $3,182.92	

unpaid invoices for Hunger Mountain Co-op. The check from this co-op has two pieces: the actual check, and also a listing of my invoices and their amounts. I find the right invoices for that check, make sure they all match, and, if so, staple the invoices to the check payment list. This bundle of invoices now has a record of payment and is placed under the deposit title page.

If the payment is a regular check without a separate listing, simply staple the corresponding unpaid invoices together and write on the top invoice, "paid check #____, $____," and add the date. If the check is for just one invoice, write the same information on the single invoice (if it's a two-part check, staple it).

All the invoices with record of payment are placed under the deposit title page. Also placed under the title page are the farmers' market inventory sheet that lists the total cash and total checks collected, and something explaining the CSA payment (a form, invoice, or note). Miscellaneous sales have invoices or a note on the title page explaining the transaction. Staple or place a rubber band around all the papers.

The bundled deposit with title page is now done and placed in chronological order with other deposits in my desk drawer.

Now for the money. Take all your checks and stamp them with a FOR DEPOSIT ONLY rubber stamp from your bank and list them on a bank deposit slip. Add in cash and calculate the total for the deposit slip. This total must be identical to your bundled deposit with title page. If not, redo your math until the two totals match. All that's left to do is take a trip to the bank to make the deposit.

Potential Financial Leaks

Financial leaks drain money from your farm business, often without you being aware it is happening. Not a good thing. Plugging up any financial leaks will keep your hard-earned money in your coffers, where it belongs. Leaks can originate from various sales avenues, or with expenses from overpayment of taxes. First, a look at the sales leaks.

Sales Leaks

CSA Programs

Most farms that offer a CSA program will print a handout describing the CSA concept. In this handout is usually a form to fill out for people who want to join. Member name and contact information is customary, along with size of the share and payment amount. When a CSA farmer receives the form with payment, a note is made on the form that records how much was paid, how it was paid, and any outstanding balance. The information from the forms is then compiled into a list of all the CSA members. It is easy to see which members have paid in full and which ones have not. If any CSA members' payments are overdue, write them an invoice for the outstanding balance. Give them one copy and keep the other for your records. This will become part of the invoice paper trail discussed above. The actual payment (by check or cash) is placed with other checks and cash to be deposited into the bank. This procedure ensures that all members are paid in full, but you need to remember to check your CSA member list for payments still due. The biggest potential for financial leaks with a CSA, though, is not payment collection, but rather giving away too much product for the money you receive. This can seriously erode your profit potential. See "CSA Potholes" in chapter 5.

Farmers' Markets

Sales income from a farmers' market is counted out from the sales apron or cash box after each market day. Start-up cash for the next market is subtracted out and left in the apron. The remaining cash and checks to be deposited are placed with other moneys to be deposited in the upright paper holder for payments, the same as with the CSA, above. Total sales for the market are recorded on the farmers' market inventory sheet for the day, and they will be reconciled with the different crops at some point (see chapter 3). All product leaving the farm through a farmers' market is paid for as it leaves the booth. There's no need to write any invoices to serve as IOUs as long as everyone who came to market pays before leaving. No financial leaks here.

One financial leak, however, is giving good weight to customers. Rounding produce weights with a hanging scale (1⅛ pound to 1 pound, for example) adds up over the course of a market, especially with high-price-per-pound items. We switched from hanging scales to self-calculating digital platform scales for easier math calculations. Our minds no longer warped adding customer totals: ⅔ pound of carrots at $2/pound plus ¾ pound of beets at $1.50/pound plus ⅝ pound tomatoes at $3.25/pound. Do that for four hours and then try saying your phone number backward. Ouch. But the real benefit of the new scales was fair trade for the farm and our customers alike. Beside easier math for the vendor, the self-calculating digital scales accurately weighed and calculated the exact price for the product on the scale. Vendor and customer could each see the weight and the dollar amount on the scale's display, and the customer was assured that human error had been eliminated. Money is money after all, and we all strive for fair trade. Moreover, the scales quickly paid for themselves by giving exact weight instead of good (or very good) weight. Here's an example:

If I sell 400 pounds of tomatoes at $3/pound, I should receive $1,200, and justly so. With my hanging scale, I'd always err on the side of the customer and give good weight. Say I made 355 equal tomato transactions of 11/8 pounds each but charged the customer for only 1 pound. I'd receive only $1,065 for the same 400 pounds of tomatoes. That's $135 less than when I use the digital scale. *And that is just one week of one item.* A $400 self-calculating scale can pay for itself in a very short time, possibly in one or two markets. And there is no downside. Customers won't complain about not getting undercharged anymore if they weren't aware of being undercharged in the first place.

Invoiced Sales

Surprisingly, sales that employ invoices can pose more of a financial leak to your business than the CSA or farmers' market, despite the reliable paper trail associated with invoices. Why so? First, invoice sales can represent a large portion of a farm's total sales. These sales are often delivered while on the road and away from your desk and are susceptible to being forgotten, altered, or lost. Math errors are more likely if you are in a hurry when making invoices at the point of delivery. CSA and farmers' market lists, by contrast, are usually tabulated at the calm of your office desk.

With invoice sales, how does money leak from your business? I presume that duplicate invoices are used. Both copies are signed by the buyer at time of delivery, and a copy is retained by the buyer and the delivery person. So far so good. But a few things can happen before you get to the bank.

No Invoice. In the beginning of this chapter, I gave an example of a community member who came to the farm and bought a bag of carrots from me. The buyer forgot her checkbook and I trusted she would send me a check. This is a potential leak that is hard to find later on: No invoice was written for the product, whether paid or not. The bag of carrots has no paper trail if there is no check (or cash and a note). Get in the habit of writing an invoice for any sale that has any possibility of slipping through your financial cracks. As mentioned earlier, I also write down (with an invoice or in a logbook in my office) nonproduce items that leave the farm, like books and tools that I loan out.

Lost Invoice. Upon delivery, a duplicate invoice is signed and a copy is retained by the buyer and the delivery person. If payment does not accompany delivery, most buyers are responsible and will pay invoices on a regular schedule without prompting from you. But what if buyers lose the invoice? They can't pay you if they don't know that they owe you.

Here's where your signed copy comes in. You bring it home to your office and place it in an accordion file holding area with other invoices awaiting payment, as described earlier. When you make a deposit, you pull out the invoices that match up with the check or payment you received. Invoices remaining in the accordion folder are still unpaid. If you come across an invoice that is overdue, you can request payment from the buyer. But what if *you* lose your copy of the invoice? It may fall through the truck's bench seat or get forgotten in a pant pocket and laundered. Without your copy of the invoice, you don't know that you aren't getting paid for delivered product. Only if the buyer dutifully sends you a check will you become aware of a missing invoice. But as a purveyor of products, the responsibility is yours to follow through on the paper trail.

There's a simple remedy to prevent invoices being lost on the delivery route. Instead of using duplicate invoices, use triplicate. With three copies of the same invoice, one copy remains in your office, and the other two take the delivery run as described above. The office copy is a backup in case the signed delivery copy gets eaten by your dog or some such loss. If the delivery invoice does make it back to your office as normal, staple the delivery and office invoices together and treat them as one for payment processing.

Should you use triplicate invoices? That depends on how many deliveries you do and how reliable your delivery people are. Triplicate invoices cost a little more and are one more piece of paper to manufacture and process. Evaluate your own system, and remember, losing money you didn't know you had is still losing money.

Math Errors

I like math and still I make plenty of mistakes. Last-minute invoice changes, hectic delivery settings, long columns of figures to add up, and interruptions can crack the best invoice writers. Miscalculations are often not caught by every buyer. Checks are written for the amount totaled on the bottom of the invoice. When you receive that check, it is matched up to the invoice or invoices associated with the check and then the money is deposited in the bank. The moral of this story is that bad math can easily go unnoticed. Some people think that is fine as long as the math error is in their favor. But honesty should prevail—fair trade is the common goal.

Two procedures address this financial leak. First, double-check your math, or have someone else do it for you. This is very simple to do and not that time consuming. The second way is to input the invoices into a computerized bookkeeping system. The bookkeeping software will alert you to any math errors so you can correct them. I've found quite a few errors this way. I then notify the buyer of the incorrect invoice so it can be rectified.

Nonsales Leaks

Taxes can take a bite out of your earnings. In discussing purchases earlier in this chapter, I showed how important it is to offset your farm income with every true farm expense because otherwise you'll be paying more in taxes. Self-employment taxes (both sides of Social Security and Medicare) and income taxes (federal and state) can take a full one-third of your net profit. Neglecting to record all farm expenses represents an unnecessary leak, but one that is easy to plug up. Make sure all your farm expenses are documented.

Another leak I recently discovered also has to do with taxes, but in a different way. We used to make most of our farm deliveries with a truck or van, expensing the costs to own and run them. We would occasionally use our personal car for deliveries, but never thought much about it. Recently, our station wagon has been used more and more to transport farm products. We kept a mileage log in the car to document our delivery trips and were amazed at how the miles added up. In 2008, we put on 2,356 miles for business-related travel. The IRS will give you a tax deduction for the mileage you put on your car—that year the standard mileage rate was a substantial $.50 per mile. That equals $1,178 in tax deductions! We feel that we can own, maintain, and drive a car for under the $.50 per mile the IRS allows. I'm not about to argue with the IRS on this one. Keep track of your car mileage and take the tax deduction.

The End: Annual Cleaning and Storing of Papers

A chapter on paper flows wouldn't be complete without addressing what to do with all of these papers at the end of the year. Sometime in mid- to late January after all the previous year's bills are paid, gather last year's paid bills folder, bank statements, and bank deposits (of paid invoices and other payments), and put them all in a large paper bag for long-term storage. Write on the bag in big letters what is inside and for what year. The IRS wants you to keep tax-supporting records for at least three years (four years for employer records). I keep these paper bags full of records in my house attic, where they are out of the way, but still retrievable in case I need to find something like a phone number or address from an old paid bill.

– 8 –

How to Retire *on* Your Farm: Retirement 101 and Business Spending Tactics

Money, money, money. All of us have different thoughts *about* money. Each of us has our own relationship *with* money. The unconscious baggage that surrounds the subject of money is often surprising and may dictate our behavior in sometimes unforeseen ways. For instance, financial security, or lack of it, can trigger deep emotions. Financial wealth is frequently confused with self-esteem and status, and financial ruin can be the instigator of one of the ultimate acts of desperation: suicide.

Why does the state of our finances carry such weight?

In our society, the dominant paradigm says that the wealthier you are, the better you are. Rightly or wrongly, we tend to elevate people who have "made it" financially. Bigger salaries are awarded to those who achieve more success. Company executives, rock stars, and professional athletes are idolized partly for their earnings. Even a pauper who hits a lottery jackpot is envied. Our culture focuses a lot more on well-paid people like Tiger Woods, Bill Gates, and Madonna than on individuals like Mother Teresa.

Why is having or making lots of money deemed desirable? Because we all need *some* money to survive, to pay for the basic needs of food, clothing, and shelter. So more money must mean that it will be easier to survive. More money allows us to have more options, more creature comforts; it enables us to get a step ahead of mere survival mode. In essence, money equals freedom. And the wealthy newsmakers like Bill Gates represent the pinnacle of this: total freedom.

That's all fine and dandy except that the corollary to this scenario says that the less you make, the less worthy you are. Employees who have lost their jobs feel this, as do those whose net worth has been devalued by a stock market meltdown. Low-paying jobs are relegated to those who are deemed "less successful." And because only a very select few can become a rock star or Bill Gates, the message is that the low-paying jobs are ones that *anyone* can do, and so they are consequently less respected.

This is not true, of course, despite what our culture thinks. Some lower-paying jobs like child care contribute to *raising our children*. What job could be any *more* important? Lifeguards at beaches and swimming pools are entrusted with our *lives* yet are paid miminal wages. Are people who prepare our food or clean our hospitals less worthy than an auto company CEO?

Farmers are businesspeople, and running a business has its ups and downs. Some years are great, some are bad. But regardless of what happens financially, our self-esteem should not be affected. Your financial *net worth* should not affect your own *self-worth,* but unfortunately, more often than not it does. You must learn to disconnect the two.

A new hobby of mine is playing tennis. In Vermont, we have a slow season, and tennis gets me off the farm; it's social, it's great exercise for the body, and it focuses the mind. I play games with friends and, in a more competitive spirit, in tournaments. As with most sports, in these competitions there is a winner and a loser. Emotions often accompany the outcome of a match: joy with winning and sadness with losing. I used to feel dejected when I lost an "important" competition match—just horrible. It was as if I felt less of myself, a lower self-esteem. How crazy is that? Why should the outcome of one tennis match affect my self-worth? Unconscious patterning dictated my feelings. I couldn't disconnect the loss on the court with the loss of self-worth. Since realizing that, I've learned to play my best tennis while remembering that it is only a game. No matter who wins, I shake hands and walk away with my ego intact.

Unconscious patterning affects our behavior without us knowing the cause. Whether your self-esteem is affected by money or a tennis match, the key to resolution is making the pattern transparent, or conscious. For a long time, I had conflicting beliefs, or a cognitive dissonance, around wealth. I resented people who were wealthy because it felt unfair that I could work so hard and not be wealthy myself. But with this same resentment came the desire to be wealthy and be one of the people I despised. That's a no-win situation. Only after thinking and talking about it did I find an answer. I now realize that there will always be people wealthier than me, and probably many more in the world who are poorer. I look around at all that I *do* have: a wonderful family, all my basic needs being met, and an enjoyable livelihood that makes for a satisfying and meaningful life. The trick is to be content with where you are, and not rate yourself against others.

Look at your feelings about money. What is your first memory of money? Take thirty seconds right now and think about it. Do these memories give you insight on how you feel about money now? As I've said before: You, and only you, are responsible for your money. Be at peace with it.

Retirement 101

I've framed this chapter around the subject of retirement, but it will cover so much more than bothering your kids from a rocking chair on the porch or playing golf in Florida. The need and strategies for retirement, saving, investing, business spending, and tax avoidance are all part of the retirement puzzle.

Why plan for retirement? For one, I'm not getting any younger, and let's face it—neither are you. Depending on your age, retirement may seem so far off that it isn't a concern, or it may be so close that you have given up hope. As self-employed farmers, we don't have another company to provide us with a retirement plan. Lifelong contributions into Social Security will provide some income, but most people would agree that it is inadequate as a sole source of income. Partial retirement or stopping work altogether means that extra money will need to come from somewhere. Regardless of what your feelings about money are, I'll assume that you'll want enough money to satisfy your basic needs of food, clothing, and shelter, and then some. Working until you drop dead is an option as long as your health holds out. But most of us wouldn't mind easing up a little as we get older. Farmers need to take responsibility for their own financial destiny. And the sooner the better, no matter what your age.

A farm needs to generate wealth in order to survive, ideally using the natural capital that we discussed in chapter 1: free sunlight, rain, nitrogen, carbon dioxide, oxygen, a living soil, and a healthy ecosystem. After covering basic living expenses, farm profit can be used in two ways: spending it or saving it.

I used to think that the best place to put my money was back into my profitable farm, but now I'm not so sure. First, if my farm was truly the best place to invest my profits, then I would be generating even more profits. Then I would have the same problem, unless I were to continually expand: a vicious cycle of making too much money. (I hate it when that happens.) Second, if I ever wanted to use my invested profits for college tuition or retirement, I would have to sell the farm, or parts of it. In reality, most farmers use this option: They work to pay off the farm and then sell it to retire. Personally, I'd like to retire *on my farm,* not off it.

I jest that my farm is not the best place to invest profits. I'll talk more about investing profits later in this chapter under "Business Spending Tactics and Taxes." But before that, I want to talk about the topic of saving.

Saving

If you plan to stop working and want to have money in retirement, where will the money come from? I list a few options as possibilities:

1. Marry someone with lots of money, or inherit it. A nice option if you can get it, but this is less under your control than the next two.
2. Save enough in a lump sum, retire, and spend it all, using your last nickel on the day you die. This works, but your timing has to be impeccable.
3. Save enough in a lump sum that it generates income or dividends, and live off the income. The principal, or nest egg, can be passed on to others after you die, or used in the last years of your life (your timing doesn't need to be as precise as in option 2).

Option 1 is the outcome of fateful circumstances, whereas options 2 and 3 involve the conscious effort of *saving* money. To save money, you must spend less than you bring in. There is no way around it. We must live within our means if we want to have money left over. Some people are natural squirrels, putting away extra for a later date. Others live hand-to-mouth, spending any money that ends up in their checkbook. But no matter where you are on that continuum, and regardless of how much money you take in, the inescapable fact is that saving requires spending less than you make. Remember *Profit = Income – Expenses*? The equation applies to your farm business as well as your personal savings.

As discussed earlier, your relationship with money factors into your ability to save money. How are you at saving money? Are you happy with your answer? If not, take time to look at why that may be. Unconscious patterns can be brought to light and habits can be changed.

Monitoring expenses is as important to saving as keeping track of income. Ultimately, spending must be reined in below your income level. The less you spend in relation to income, no matter how great or little the income is, the higher the savings. How can spending be kept in check? Here are some suggestions:

- **Don't buy things you can't afford.** Ask yourself if you really need the item. I advocate fiscal prudence, especially when starting up a farm. Think of your wallet as a clam that you have to pry open to spend money. Watch for spending "black holes": items for which you pop open your wallet without thinking. It is surprising how much money black holes can absorb without you noticing: music CDs, clothing, computer gear, eating out, books, or farm machinery. In regard to the last item, a common affliction is "new paint disease," an expensive habit. For me, a recent black hole is tennis. I first replaced an antique racket with a newer model, then I started taking lessons, joined a tennis club in the winter, bought better tennis shoes, went to weekend tennis camps, and emptied the bookstore shelves of tennis books. It's like my rational mind goes into hibernation where tennis-related purchases are concerned. I'm not saying we shouldn't spend money on fun things, for we all need to enjoy the fruits of our labor. Just be aware of the amount of money that goes to different places. *Saving money means not spending as much money as you bring in.*

- **Be efficient.** Buy in bulk, and search for the best deals. Communicate with other farmers to get the latest information and coordinate group purchases. *A dollar saved is $1.34 earned.* Because of the tax bite on income, you need to make $134 to be able to spend $100 on nonfarm purchases.
- **Pay off debt intelligently.** If you have $1,000 to pay down debt, pay off the highest interest rate first, if possible. Often that means credit card debt. Imagine your checkbook as a bathtub half full of water. The water in the tub represents the money you have. Money flows into the tub through the faucet (your income plus any savings account interest) and out the drain (your expenses and debt interest payments). Now imagine that your income and expense streams are exactly the same; all that's left is the interest income dripping in and debt interest payments dripping out. If you are making 2 percent interest income on your savings and paying 18 percent on your credit card, that's like two drops of water coming into the tub and eighteen drops leaking out. What happens to the level of water in the bathtub? It will slowly be drained off. Your water level will stabilize if you forsake your two drops of savings in order to stop the eighteen drops leaking out.

Raising income may be an option to increase savings, but all too often increased spending follows suit. Saving can occur at any income level as long as basic living needs are being met.

Investing

Once you have saved up some money, what do you do with it? Here are a few possibilities:

1. Stuff cash under your mattress.
2. Buy land.
3. Purchase baseball trading cards.
4. Invest in stocks and bonds.

Consider the options.

1. Cash is accepted everywhere (it is a very *liquid* asset), but it loses value over time due to inflation. A dollar

today buys more than a dollar will ten years from now. With cash under your mattress and a 3 percent inflation rate, you lose 3 percent of your buying power each year. You own the same amount of dollars, but they buy less. The longer you hold on to cash, the bigger bite inflation takes from your purchasing power.

2. Unlike cash, land is not a very liquid asset. Selling land takes time and some legal work if you hope to convert it into cash that you can spend. Land is real estate, and most people resonate with the *real*ness of it, but the value of land fluctuates. Trends show land value increasing over the long haul, making more than inflation, but not for every location or block of time.
3. The same trend holds true for the baseball trading card investment option. Over time, and with some knowledge or luck, certain baseball cards may pay off. I include this oddball option of collectibles somewhat lightheartedly as a precursor to the next choice, stocks and bonds.
4. Stocks and bonds have been an investor's traditional option for many years. The subject is huge; I'll only touch upon it to make some points on investing and retirement. To learn more, search out more information via books, magazines, websites, or investment professionals.

When buying stock in a company, you own a percentage of the business. Your invested money in the company's stock is tied to the success or failure of the company's business. The value of your investment will fluctuate accordingly. Bonds, on the other hand, are a loan. When you purchase a bond, you lend money to a company, a municipality, or even the US government (Treasury bonds). These loans are paid back with interest, and your investment value doesn't fluctuate as it does with stocks. In general, bonds are less risky than stocks. In the long run over the last seven to eight decades, stocks have averaged about an 8 percent return on investment; bonds, about 5 percent.

Many farmers have a knee-jerk reaction to stocks and bonds. These investments are the epitome of paper dollars: profits made by buying and selling, each trade with a requisite winner and loser. Farmers in contrast deal in solar dollars and generate a real and tangible product. Stocks and bonds are often associated with wealthy investors, not farmers, which creates an inherent bias. I understand this concern, and in no way am I promoting stocks and bonds as an investment vehicle. But to be at peace with the subject, I'll offer the following thoughts.

Like nature, our current economic world is made up of an extremely complex and diverse series of interrelationships. Each organism or business finds its own niche to survive. Some thrive more than others. Some fail. When a niche changes, others adjust to the new landscape. As farmers, we are tied into this business web whether we admit it or not. Nary a farmer is clean of the ties to corporate America. We support companies through our purchases without actually buying stock in the company. Buying fuel for our truck, batteries for our tractor, and even the tractor itself all supports the companies that make the products. Unless we return to horse-and-buggy days, we all support big companies that we may not readily invest in. Many products we buy don't come from small cottage industries; they are from large-scale, complex businesses. Manufacturing tractors is no small feat; diesel engines, synchronized transmissions, and rubber tires have been years in development and required the concerted effort of many people.

Socialism, communism, and capitalism all have their faults and detractors. We in the United States are rooted in capitalism. I was born in the 1950s, and growing up in my household the word *profit* was equated with *greed.* I wasn't aware then of a "good" profit. Communist or capitalist, greed is the ugly human behavior that crosses all boundaries and is responsible for many of the world's miseries. In pursuit of profits, many a company has let greed trump moral behavior. This doesn't have to be the case. I'm a believer in *good* capitalism, one where a business can profit *and* do good.

Our purchases are an act of intention. We can choose to buy one product over another or to buy no product at all. We decide how our money is spent. Investing in stocks and bonds is no different. I can buy stock in a company that makes solar hot-water panels or one that makes computer software. I can buy a bond for my local co-op to fund an expansion or one to help fund the US government. My message is that we are all interconnected and complicit in the economic world, and we can choose how to interact with it.

An Investment Primer

Principles of sound investing are simple and commonsense:

1. Make a plan and follow it.
2. Buy good-quality investments, confirmed by an independent source.
3. Buy low, sell high.
4. Diversify.
5. Start early and be in it for the long haul.
6. Be tax-smart. Contribute the maximum amount to your IRAs.

Investments can be made in anything that appreciates in value over time. This may be your farm, an acre of young walnut trees, a truckload of copper plumbing pipe, baseball cards, or financial instruments like stocks, bonds, and CDs. Whatever investment you make, it should follow a plan of what your financial goals are and how and when you want to achieve them. When will you retire? How much money are you aiming for in retirement?

Are there big expenses in the future like a land purchase or children's college tuition? How much do you need to save each year? How much of an emergency fund do you need? What will projected expenses in retirement be? Think over your goals and consult the numerous resources available to create a plan, and then remember to follow it.

Because the topic of financial instruments is rarely the farmer's dinner-table talk, I'll outline some common investment choices:

1. **Stocks:** Partial ownership in businesses—small to large, domestic or international, risky or less risky.
2. **Bonds:** Loans to businesses, governments, schools, or municipalities. Instead of borrowing money from a bank, organizations issue a bond and repay it with interest.
3. **Mutual funds:** A collection of different stocks and/or bonds that a professional manager oversees for a small fee. Mutual funds can diversify your investment portfolio.
4. **Real estate.**
5. **Annuities:** A contract backed by an insurance company.
6. **Miscellaneous:** Collectors' items like baseball cards and precious metals like gold or silver.
7. **Certificate of Deposit (CD):** Basically a loan to a bank. A safer but lower-yielding investment than stocks, bonds, or mutual funds of combined stocks and bonds.
8. **Money market account:** Like a higher-yielding checking account, but check writing often has a minimum amount requirement, such as $500.
9. **Savings account:** Offered by banks; safe, but with low yields.

Most of the investment options listed above fall into two main groups: growth and savings. Growth investments are stocks (and mutual funds composed largely of stocks) and real estate; these investments pose more risk than others but also offer greater reward. Investors with a longer time until retirement (and who are able to ride out any short-term ups and downs of the market) benefit from this group with its increased growth potential. In contrast, savings investments are bonds, CDs, money market accounts, and savings accounts; these are less risky than growth investments and consequently yield less. CDs, savings accounts, and some money market accounts are insured by the Federal Deposit Insurance Corporation (FDIC) for up to $250,000. If your bank or financial institution fails, your money is guaranteed by the US government. While savings investments are safe, their annual percentage growth may be only 1 to 5 percent.

Business Spending Tactics and Taxes

Investing in your farm is always an option for money you have saved. I joked earlier that this may cause a vicious cycle of making too much money *unless you continually expand your business*. When your farm generates a profit, you can save the money and possibly invest it off the farm. Or you can spend it on farm items and services. The vicious cycle of making too much money is broken when you take farm profits and reinvest them in your farm. In expanding your business, your farm profit is reduced by increased farm spending. Your annual income or farm profit is lowered, but your net worth (the value of everything you now own) is increased.

The IRS tax laws encourage businesses to capitalize their operations through *tax avoidance*. Tax avoidance is different from tax evasion. The first is legal, the second is not. If a farmer makes $36,000/year net profit, one-third of that amount ($12,000) will go to paying taxes (15 percent federal income tax, 15 percent self-employment tax, and about 4 percent state income tax). After paying taxes, only $24,000 is left for the farmer. Ouch.

But if before year end the farmer buys a tractor for $12,000 and writes off the entire expense in a year, the farm's net profit would be only $24,000 and the taxes on that would be only $8,000. That's a tax savings of $4,000. *It's as if the $12,000 tractor cost only $8,000!* Showing this in table form (table 8-1) helps illustrate the point.

In avoiding taxes with the tractor purchase, the farmer has less after-tax income but still has the value of a $12,000 tractor, coming out ahead of the scenario that involves paying taxes on all of the $36,000 net profit. This example shows how tax laws encourage businesses to invest in themselves.

Tax avoidance is the main reason I capitalize my farm business. If I have a less profitable year and I'm not in need of a tax advantage, any machinery I purchase must strongly increase farm profitability; that is, the machinery must pay for itself quickly by saving other costs. If I have a banner year, however, I will look for

TABLE 8-1

	No tractor	With $12,000 tractor purchase
Farm net profit	$36,000	$24,000
Minus taxes on net profit	-12,000	-8,000
Profit left for farmer	$24,000	$16,000
Add value of tractor	+ 0	+12,000
Net worth of scenario after taxes	$24,000	$28,000

ways to effectively buy down my projected net profit before year end. But that doesn't mean I'll spend profits carelessly.

I caution farmers to be careful about their capital purchases and guard against "new paint disease" or machinery proliferation for a couple of reasons. First, machines may not have brains, but they tend to make decisions for you. How so?

Let's say I grow beets for a number of years and get to a scale where harvesting costs quite a bit in labor. I buy a specialized beet harvesting machine for $10,000, saving me money each year in harvesting expenses. But now with money tied up in a harvester, I'm much more likely to continue to grow beets on this scale or larger, and I'm inclined to grow other crops to make use of my harvester. Whether those decisions are financially sound is not the message. The point is that the machine is, to some degree, making decisions for me and influencing my farming practices. That may be fine, but it's important to be aware that it is occurring.

The second reason I warn against excessive equipment purchases is that *things that rust, rot, and depreciate are not investments, they are expenses*—a saying coined by England's Dr. Gordon Hazard. Investments should maintain or increase in value over time. A new manure spreader is not an investment and never will be, and neither is a tractor. Both will lose value over time and require maintenance to boot. A manure spreader may be a necessary tool for the farm, but it is not an investment. If you truly need a manure spreader for your operation and need to buy down farm profits for tax avoidance, fine. But remember: It is an expense, not an investment. When in doubt, keep in mind that investments need to *increase* in value.

Tying It All Together

Back to Retirement 101. How do these concepts of saving, investing, business spending tactics, and taxes come together for retirement planning? I find it helpful to show some very simplified examples of different scenarios over varying time periods. All four examples show a projected net profit of $36,000 at year end. To illustrate my points as clearly as possible, I do not account for the tax benefits of IRAs, the effect of inflation, or the need for living expenses.

Option 1A

I buy nothing before year end, pay my third in taxes on $36,000 net profit, and end up with $24,000 net income after taxes. I put the money in my checking account, where it earns 0 percent interest.

Option 1B

Same as above, except I put the net income after taxes ($24,000) into a stock mutual fund that averages 8 percent return.

Option 2A

I buy a new $24,000 tractor before year end and expense it all in the tax year. My net profit is now only $12,000 after buying the tractor. I pay my third in taxes on the $12,000, leaving me $8,000 in after-tax net income. I put this money into my checking account, where it earns 0 percent interest.

Option 2B

Same as Option 2A, except I put my net income after taxes ($8,000) into a stock mutual fund that averages an 8 percent return.

How do the different options look at year end and after ten years, twenty years, and thirty years? Table 8-2 shows a comparison.

Look at the different options after thirty years. What a difference in outcomes! Eight thousand dollars to a whopping $262,458! Who would have guessed?

Remember that this is a simplified example, but it is not necessarily unrealistic. It represents one year of investing only, not the more likely scenario of the multiple smaller investments we might make in real life. But the 8 percent return from the mutual fund snowballs each year, building on itself. I'm not pushing mutual funds; any investment that maintains a steady growth will work. A very safe CD or US Treasury bond at 5 percent will yield $107,226 in Option 1B, far less than the 8 percent return of $262,458, but still a significant increase on the original $24,000 investment.

With retirement investments, I can't overemphasize *starting early*. Time is one of your biggest friends in investments, whether it is with stocks and bonds or an acre of black walnut saplings. It is never too late to start. Begin

TABLE 8-2

	Year End	10 Years	20 Years	30 Years
Option 1A				
Checking	$24,000	$24,000	$24,000	$24,000
Option 1B				
Mutual fund	24,000	53,271	118,243	262,458
Option 2A				
Checking	8,000	8,000	8,000	8,000
Tractor value	+24,000	+12,000	+6,000	+0
Net value	32,000	20,000	14,000	8,000
Option 2B				
Mutual fund	+8,000	+17,757	+39,414	+87,486
Tractor value	24,000	12,000	6,000	0
Net value	32,000	29,757	45,414	87,486

by first making a plan, then start saving some money and investing it wisely. Remember to maximize your IRA contributions each year for tax-efficient investing. The tax savings can be dramatic.

The take-home messages from these different options are:

- Machinery purchases can be advantageous in reducing taxes on net profit but are expenses, not investments.
- Saving and investing for retirement snowballs to a much larger amount over time. Investing early for retirement is better than getting a late start.
- Taxes take a large bite (one-third) out of your farm earnings. Contribute the maximum amount to your IRAs each year if possible. (Also see the section on SEP IRAs in chapter 6.)

As a disclaimer, I'm a farmer and not a tax or investment adviser. I want to give you a basic idea of the factors that affect retirement because I feel that we all need to be responsible for our own financial destiny. I'm fully aware that young people who benefit the most from early saving and investing are also the least able to salt away money when building a farm business. Do what you can. My thoughts are just a starting point. Educate yourself, make a plan, and try to stick to it. Some good books on the subject are *The Only Investment Guide You'll Ever Need* by Andrew Tobias and *Personal Finance for Dummies* and *Investing for Dummies,* both by Eric Tyson.

– 9 –
Production Efficiencies

On the first page of chapter 1, I noted that there are already numerous resources available on the production aspects of farming. Instead of adding more to those resources, I wanted to share my thoughts on the often neglected business end of farming. So why then a chapter on *production* efficiencies? Because in our old equation Profit = Income – Expenses, one of the ways to increase profit is by lowering expenses. And how can you lower expenses? By being more efficient.

In working with numerous other farmers, I see over and over again a lack of production efficiency. Farming is a production business; farmers get paid by the piece (by the watermelon or head of cabbage sold), not by an hourly wage. So it's crucial to focus on making every aspect of the farm business as efficient and productive as possible. To realize profits for any given crop, economy of effort is paramount. If an employee can do a task in a shorter amount of time, farm expenses are reduced and profit increases. Because organic farming can be a labor-intensive business, efficiencies in labor usually yield the biggest impact on profit. Of the labor needed for most crops, the biggest rewards come first from more efficient weed control, followed usually by harvesting and pack-out. To address these production efficiencies and others, I'll describe systems I've found to work, ones that I use to this day.

I have a tractor and various implements for field preparation, cultivation, and some harvesting. I also use many hand tools. Even if you don't own a tractor or any implements, though, the goals of efficient production are similar: good yields with less effort. Most of the implements that I describe below require the use of a medium-size tractor, in the 35- to 50-horsepower range. I also use a smaller 1950s cultivating tractor with a whopping 10-horsepower engine. Cultivating with a larger tractor works fine, but I find the smaller tractor to be more precise in eliminating weeds. The first part of this chapter focuses on efficient weed control, which begins when you first till the soil, long before the weeds even germinate. The last part of the chapter talks about efficient seeding, harvesting, and greenhouse strategies.

Standardize and Raise Your Beds

Adopting a standardized row spacing for your whole farm is one of the simplest ways to increase efficiency. I use raised beds spaced 6 feet on center apart, with three rows per bed. Rows in each bed are spaced 14 inches apart, perfectly parallel to each other. If you use a tractor to form the beds, wheel spacing should be set to 60 inches (tread center to tread center). All tractors on the farm should be set to the same wheel spacing. There will be roughly 46 inches between the inside of the rear tires, but this varies a little depending on the width of the rear tires. If three rows are spaced 14 inches apart on the bed, the two outer rows are 28 inches apart from each other, and 9 inches away from the inside of the rear tractor tire. The 9-inch space from the outer rows to the tires is needed as the plants grow in size and as a margin for error if you are cultivating with a tractor.

Plant all crops on this universal three-rows-per-bed system. Smaller plants like carrots, beets, lettuce, and spinach are planted in all three rows of the bed (14 inches apart). Crops that need a little more room, like peppers and eggplant, use just the two outer rows (28 inches apart). Large plants, such as melons, squash, cucumbers, and artichokes, occupy the middle row only (6-foot spacing overall).

The purpose and overall benefit of a standardized bed system is uniformity in all aspects of growing the crop. All tractors are able to straddle every bed, at any time, for planting, transplanting, fertilizing, weeding, spraying, mowing, and harvesting. Cultivating tools can thus go over any crop with minimal to no adjustment. Pickup trucks also are able to drive over any bed on the farm for quick transport of some harvested crops.

Tractor cultivation for weed control is the biggest benefit of standard row spacing, and

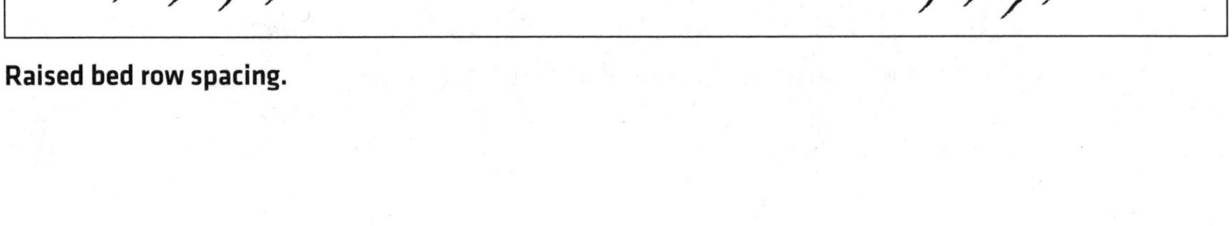

Raised bed row spacing.

Two- and three-row beds.

weed control is often the biggest expense in a crop budget analysis. Hand weeding may take two hundred hours or more per acre, even with mechanical cultivation for weeds. Precise tractor cultivation saves hours and hours of hand weeding for all crops. To effectively cultivate weeds later in the season, certain steps need to be taken from the outset of bed preparation. I'll discuss different types of tractor cultivators in detail later on in this chapter. First, though, I'll show my system of how beds are made.

The purpose of making a bed is to provide:

- A deep, level, aerated, and finely textured seedbed to seed or transplant into.
- Good air and water drainage—important for plant health, and thus disease and pest control.
- Uniform spacing for ease of weed control and harvesting.

Additionally, raised beds are less prone to wind erosion than flat bed culture, and the level, tabletop surface aids in mechanical cultivation.

Making Beds

Beds can be made with a shovel and rake by hand, a walk-behind rototiller, a draft animal, or a tractor and attached implement. Once you start growing more than an acre of crops, mechanized bedforming and weed control become cost-effective. My process for making beds with a tractor and various implements will give you the overall picture; then I'll talk about weed control.

When preparing a field to plant, I follow soil test recommendations and spread any compost or organic fertilizers that are needed. A manure spreader applies compost, and a cone spinner spreader applies any granular rock powders or fertilizers. If the field has a cover crop that needs to be incorporated, I'll use a disk harrow to incorporate the plant material and amendments that I spread. I own a moldboard plow but use it only to turn in thick sod. If the field is fallow and without much plant residue, I'll use either the disk harrow, an S-tine harrow, or even a chain harrow. Disking is better for

Disk harrows.

S-tine harrows.

Chain harrows.

incorporating lots of plant material into the soil. The S-tine and chain harrows incorporate soil amendments sufficiently in more open soils and use less horsepower to pull.

After the field is fertilized, I start forming the beds with a mini chisel plow. The chisel plow rips and loosens the soil with three 14-inch-long spring-loaded shanks spaced 14 inches apart. Attached to the rear of the implement are two hilling disks to move soil from the wheel tracks into the bed area. I added on these disks to help with the final bedforming. Chiseling aerates the soil at least 10 inches directly below the row to be planted without inverting the soil. It is a joy to watch from the tractor seat. Unlike rototilling, chiseling doesn't create a "plow pan" (a hard layer beneath the plow or rototiller tines); nor does it beat air into the soil, which speeds up organic matter breakdown. For deeply rooted crops, I may chisel a bed twice. Mini chisels are smaller than their larger cousins, which have longer shanks and require more horsepower to pull. Mini chisels are better suited to making individual raised beds.

I can't overstate how much I like this mini chiseling. Growing conditions in sandy or heavy soils improve drastically once the soils are chiseled. Older, used chisels called Ferguson field cultivators can be found for $100 to $300. New mini chisel plows from SSB Tractor (www.ssbtractor.com) cost about $700, plus shipping. I prefer to use only three shanks spaced 14 inches apart, removing the outer shanks that follow the rear tires. If there are outer shanks directly behind the rear tractor tires, the more compacted soil in the wheel tracks keeps the implement from digging as deeply as possible. This prevents the middle three shanks (the ones that really matter) from sinking all the way down within the bed. Besides, aeration in the wheel tracks is less useful and will be negated by future tractor passes.

In soils with lots of cover crop or plant residue, it is best to disk in thoroughly or wait for some decomposition to occur. Chunks of fibrous material can get jammed between the shanks and keep the implement from sinking in fully. To prevent this clogging, offset the chisel points so they don't all line up with the toolbar. Use a shank with an extended arm for the middle chisel so that chunks of plant material flow more easily between the chisel shanks.

I chisel a number of beds at one time for increased efficiency. I usually make the number of beds I'll need in the next two weeks or so. I pay particular attention to making the beds extremely straight and evenly spaced apart. I overlap the next bed's wheel track by an extra half tire width so I have a little wiggle room in the future when straddling beds with a tractor or truck. This widens the 5-foot tractor wheel spacing to a 6-foot bed spacing. To make laser-straight beds, I line up the first bed with the edge of the field or the last previously made bed. If I have to begin chiseling in the middle of an unbedded field, I pace off the distance from my last bed to determine my starting and finishing points for the bed

Mini chisels with hilling disks.

Chiseling beds.

to guarantee that all the beds will be parallel. I start the bed by focusing on a fixed object beyond the end of the bed I'm making. Keep focused on that object as you proceed down the field. Keep in mind that whenever you turn your head around to look at the implement behind the tractor, your arms and hands naturally tend to turn the steering wheel. Focus on keeping the steering wheel steady when glancing behind you. When you chisel a number of beds, straighten out any curved beds by going over the same bed again or adjusting with the next bed.

The chiseled beds should be straight, evenly spaced, aerated, and slightly raised. For the final bed preparation, I use a Larchmont-style bedformer. This tool uses hilling disks and shaping boards to pull soil from the wheel

Bedforming. Note the three springs used to mark rows.

tracks into the center of the bed and then flattens the soil with a leveling board. Near the rear of the bedformer are straight disks called meeker harrows that knife the soil to break up any clods. At the very back of the bedformer, I bolt three 6-inch coil springs spaced 14 inches apart to the rear of the frame. These springs mark three parallel rows on every bed I make. Bolts protruding from the frame also work to mark rows but may become bent without your realizing it. You can mark every bed with your rototiller in the same way, by attaching springs or bolts to the rear apron. This guarantees equally spaced and parallel rows to transplant or seed into every time you form a bed, without any extra passes or effort.

A well-formed raised bed is 4 to 6 inches or more in height from the depth of the wheel track. The top surface of the bed is flat, with no dips or divots, and the soil is firm enough to walk on without sinking in more than an inch or two. The three parallel lines left by the bedformer's rear-mounted markers should be clearly visible, even after a heavy rain.

If there are clumps of sod or fresh plant residue, care needs to be taken that the bedformer doesn't temporarily clog and leave a furrow in the bed. If this happens, raise the bedformer slightly so that any residue can pass through. Bedforming the same bed twice is usually time well spent, with or without the clogs. When turning around at the end of a bed, take care to keep the beds straight and equally spaced apart. I chisel and bedform in fourth or fifth gear (out of eight total forward speeds, roughly 3 mph). An acre takes about an hour and a half to chisel once, and about an hour and a half to bedform once.

Kill Your Rototiller

A word of caution is needed about the common practice of rototilling. While it's a very effective way to make a seedbed in one pass, many farmers greatly overuse this tool. Rototillers are easy and satisfying to use: A clean-looking, finely textured seedbed can be created relatively quickly. Continuous rototilling, however, comes with some serious drawbacks. First, the tines till only to a depth of roughly 6 inches, and, in doing so, the action of the tines creates a hard layer of soil just underneath the tilled soil—a different type of "plow pan." Second, air is whipped into the soil, which "burns up" or greatly increases the decomposition of organic matter. Third, the fluffed soil quickly settles back down to grade, undoing the benefits of a once raised bed. All this is not to say that rototilling doesn't have a place on the farm; the chronic overuse of it is what causes concern.

Fortunately, it's possible to achieve a happy medium. Preceded by a pass with a mini chisel, a rototiller spun at a slow RPM (or with a faster tractor ground speed) has fewer deleterious effects than normal rototilling. This combination creates a good, clean, slightly raised, finely textured seedbed with deeply loosened soil after only two tractor passes.

Bolt markers on rototiller apron.

Rototilled bed with three rows clearly marked.

Weed Control with Tractor Cultivation

Weeds are the bane of almost any organic farmer. The onset of a flush of weeds on a farm demands immediate attention and can be overwhelming to the farm's management and workforce. Mechanical tractor cultivation and flame weeding can hugely reduce weed pressure (and its cost) and may mean the difference between crop success and failure. I will spend a lot of time on this subject because of its importance to a farm's profitability.

Tractor cultivators can take many forms. Tools may be either rear-mounted on a three-point-hitch toolbar or attached to a belly-mounted toolbar between the front and back wheels (under the tractor's engine and transmission). Belly mounting is preferred for visibility and accuracy, especially when plants are small. In addition, belly-mounted cultivation tools are less affected by the tractor's steering than is a rear-mounted toolbar.

There are many types of cultivation tools, but three stand out as most common and useful: tine weeders, basket weeders, and sweeps.

Tine Weeders

I farmed for many years without a tine weeder, mainly because I didn't know what I was missing. Workshops at other farms that I attended demonstrated its use, but I failed to see its benefit to my operation. That all changed a few years ago.

The small-diameter metal rods of a tine weeder scratch the surface of the soil over the *entire* growing area, right on top of the growing crop as well as in between the rows. How can that work? Timing is key. The planted crop must not dislodge as easily as the weeds do. A bed with deeply planted seeds like corn or beans can be tine-weeded before the crop even emerges. And young seedlings (whether transplanted or direct-seeded) that are better rooted than the surrounding weeds survive the tines' action, while the weeds don't. Tine weeders "tickle" the soil with a light, vibrating scratching of the soil surface. Only the small weeds are uprooted and killed. Larger weeds remain, the same as a well-rooted seedling would.

The big advantage of tine weeding is that small weeds *in the row* are uprooted and killed while leaving the crop unscathed. Most other mechanical cultivators kill only weeds *between the rows*. Tine weeding is also much quicker because it is "blind" cultivation. The tractor driver doesn't need to worry about hitting the rows of crops, because the tine weeder cultivates the *entire* bed.

Tine weeders are a favorite tool of mine to eradicate weeds, whether the crop is direct-seeded or transplanted. As long as the young plant is better rooted than the germinating weeds, future weeds are eliminated, especially the ones in the middle of the row that would normally need to be hand-weeded. Broccoli and lettuce transplants can be tine-weeded after a week or so, and even direct-seeded crops like beets or parsnips when they are only 3 inches tall. Set your tine weeder to cultivate as lightly

Blind cultivation with tine weeder.

as possible at first; do a small section and check the results. It seems strange at first running tines directly over your crop, but there's no need to worry if your crop is well rooted.

Basket Weeders

Basket weeders are used anytime after beds are made (before or after planting) up until the plants grow too tall to pass unharmed beneath the basket weeder axles (up to about 12 inches high). Baskets throw very little soil side-to-side, allowing extremely close cultivation of each row of crops. I prefer 6-inch-wide and 2-inch-tall baskets, with 8-inch spacing in between each set of baskets.

The 8 inches between baskets allows for cultivation of leafy crops like lettuce, but also very close cultivation of upright or small plants. The close cultivation is possible by making two passes over the same bed. With your first pass, run the left edge of the baskets as close as possible to the row(s) of crops, often ½ to 1 inch away. After you turn around at the end of the bed, the second pass again places the left edge of the baskets close to the crop. All that remains untouched are the three narrow bands

Rolling basket cultivators, belly-mounted.

Close-up view of rolling baskets.

Close cultivation on right side of crop with rolling baskets. Rear sweeps cultivate wheel tracks.

where the crops are planted. Before I learned this technique, I would make only one pass per bed, leaving an 8-inch band to hand-weed. Now with the two-pass offset method, only 1 to 2 inches remains to be hand-weeded. *That eliminates 90 percent of the area to be hand-weeded!* All that for two passes with a tractor-mounted cultivator, which takes about two hours per acre total.

Double-pass offset cultivation can be used as the crop grows bigger, getting as close as you dare. And finally, just as the plants get too big to pass undamaged under the basket weeder's axles, make one last pass, centering the baskets between the rows of crops.

When using baskets on a crop that is planted down the single middle row of the bed (like squashes), or crops planted on the two outer rows only (like peppers), I put a sweep on the blank row(s) to kill the weeds not touched by the baskets. These sweeps are mounted on the basket weeder frame or behind the rear wheels. Note: Remember to remove the sweeps before cultivating a three-row crop!

While close cultivation is much easier on flat fields, sometimes beds are made on fields with a slight side slope. You can increase accuracy on sidehills by gently applying the brake on the uphill side of the tractor. This counters the force of gravity pulling the tractor's front wheels downslope, because tractor steering alone sometimes isn't sufficient.

Sweeps

Once the crop grows too large for the basket weeder, V-shaped sweeps are used to cultivate weeds between the rows. Sweeps move soil side-to-side, which is problematic with small, tender plants, but highly beneficial when the plants are sturdier and can withstand some hilling. When used correctly, sweeps can throw soil into the row of a plant that is 6 inches or larger and bury any small weeds *within* the row. Sweeps can also reach areas under the foliage of leafy plants without disturbing the crop itself.

Cultivating is an art as much as a science. Take your time fine-tuning your cultivation setup so it works as intended. Bring a couple of wrenches with you and be prepared to hop off and on the tractor a few times when starting out. A little time spent up front adjusting cultivating equipment will save many, many hours of hand weeding.

The width of the sweeps and the depth at

Using sweeps to throw soil in the row.

Sweeps are all in line.

which they are set affect the amount of soil cultivated. I generally use 4- or 6-inch-wide sweeps and use depth control to vary the amount of soil moved. The front points of all the sweeps must be aligned in a straight line perpendicular to the rows to move soil evenly *into* the row of plants, so the plants are propped up with soil, not buried.

I use a tape measure to set the front points of each sweep 14 inches apart. Getting the right spacing is important so that when you are on the tractor seat and focusing on one row, you are confident all the other rows are being cultivated in a similar manner.

Sweeps can be used belly-mounted or on a rear-mounted toolbar. The previously mentioned advantages of belly-mounted tools over rear-mounted ones still apply: better visibility and being centered in the turning radius. But once the crop is larger and sturdier, rear-mounted sweeps work well, too, and can be used on tractors without a belly-mount option. For example, I use rear sweeps on certain plants when they are over 4 feet tall! The plants bend under the tractor and toolbar and straighten up in a few hours.

Three-point-hitch toolbars for attaching rear sweeps need to be adjustable to accommodate

Adjusting sweeps with a tape measure.

Cultivating wheel tracks with rear toolbar.

Turnbuckle to eliminate sway in rear lift arms.

Rear-wheel track eradicator.

different crops at various stages of growth. To stabilize lateral movement of the entire toolbar, turnbuckles connect the two lower lift arms to the tractor rear axle. A toolbar with side-to-side play will reduce the precision of mechanical weed control.

Wheel tracks are notorious for germinating weeds. Every time I cultivate with belly-mounted tools I put wide (8- to 12-inch) sweeps directly behind the rear wheels so that every pass with the tractor disrupts weeds growing between beds. The same goes for rear-mounted cultivators. And often, when the plants in the bed are large, I'll use these wheel-track eradicators alone at higher speeds to clean up weeds in the open soil between beds.

Stale Bed Seeding and Flame Weeding

Stale bed seeding refers to preparing a bed and leaving it to germinate weeds, then killing the weeds either before or after planting the intended crop. The most common practice is to form the bed, let the weeds germinate, and then re-prepare the bed or lightly cultivate the soil surface to kill the weeds. Since most weed seeds germinate only in the top 2 inches of soil, many future weeds for the season can be controlled with this method. Crops can be sown just after the re-bedding or light cultivation, and they will germinate with less competition from weed seeds.

Another stale bed scenario employs a tool called a flame weeder. Flame weeders are usually

Cultivating wheel tracks while cultivating a bed.

Rear cultivator for quick cleanup of wheel tracks.

Tractor-mounted flame weeder.

Handheld flame weeder.

propane burners that are passed quickly over germinated weeds. The flame doesn't actually burn the weeds but rather kills them by breaking their cell structure.

Weeds looked "steamed" and quickly die. I liken the process to a frost kill, but using heat instead of freezing temperatures. The major benefit of flame weeding is that the soil surface is not disturbed, and thus new weed seeds are not brought to the soil surface where they can germinate. If an intended crop takes a long time to germinate (as with parsnips and carrots), then flaming the bed just before the crop emerges will kill all the weeds that *have* germinated. The planted crop surfaces a day or two later to a perfectly weed-free landscape.

Timing is everything! Flame weeding is most beneficial if done just before crop emergence. Flaming will kill the intended crop if it has already started to break through the soil. There are a few exceptions—crops that can tolerate a light flaming, including potatoes, onions, and corn—but even these will be set back some.

Flame weeders have been around since the 1930s but took a backseat to herbicides until the 1990s. Flame weeders can be handheld or hand-wheeled units that are attached to a barbecue-grill propane tank and carried around the field. Tractor flame weeders are mounted on a three-point hitch with a large propane tank. All work the same way, passing a hot flame over the bed at walking speed.

Flame weeding is ideal for crops with long germinating periods such as parsnips and carrots because many weeds germinate in a shorter time period and can be flame-killed. The technique also works with crops that germinate in a shorter period of time, but in this case you need to pay close attention to germination.

Prepare the bed to be seeded and wait one to three weeks if possible before seeding. Irrigating stale beds will hasten weed seed sprouting. Weeds germinate quickly; in fact, most are genetically programmed to do so. Plant your crop by running your seeder on the bed, following the lines left by the markers on the bedformer or rototiller. Immediately cover a few small sections of the row with a row cover, Plexiglas sheets, or glass panes.

The microclimate under these temporary covers speeds the germination of your crop (as well as any weeds) and gives a one- or two-day look into the future of the uncovered rows. *Check under the covers daily* to see if the crop has started to emerge. Try digging up a seed or two to see if it has sprouted. When the seeded crop pops through the soil surface under the cover, the rest of the crop will emerge in a day or two. This is the time to remove the covers and flame the entire planting. Don't put it off. You have a small window to kill the weeds without killing your crop. Flaming can even be done in the rain! Be vigilant in monitoring the crop under the covers and act quickly when the time comes.

Why all the fuss about flame weeding? Simple math. With a tractor-mounted flamer, two hours are required to cover an acre, using $60 of propane. Flaming reduces overall hand weeding on an acre from two hundred hours per acre to eighty or less. That's a savings of 120 hours per acre, or $1,500 in hand-weeding expenses (at $12.55/hour). Subtracting the tractor and propane costs, more than $1,400 per acre is added directly to the crop's bottom line.

Other Weed Strategies

Most farmers find that one or two weed species present more of a problem for them than others. Pay attention to those weeds' life cycles and learn as much about them as you can. Often you can adjust your farming practices to lessen the weed pressure. Using row covers or irrigation to hasten germination of weed seeds so they can be killed before planting is one technique. Applying straw or plastic mulch is another common example.

One strategy for me is to time a planting to avoid weed pressure. Galinsoga is a pernicious weed on my farm. It grows and sets seed quickly, it is hard to kill mechanically once well rooted, and its large root-ball when hand-weeded disrupts small crops like young carrots. Field fertility adjustments have been said to help, but I haven't nailed that one down yet. I prefer to adjust my crop timing.

Galinsoga is an annual weed and germinates only in warmer soils, typically in June here in Vermont. This gives me four to six weeks' lead time to seed and transplant crops to outcompete the galinsoga. Early cultivation gets the crop up and thriving before serious weed pressure develops. And since galinsoga won't tolerate a freeze, I can use that to my advantage with frost-hardy crops. An example of this is overwintering field spinach. I seed spinach in the fall in time for its leaves to grow to 3 or 4 inches long. The plants then go dormant over the winter and start their regrowth in April before any weeds even think about germinating. The spinach is ready to pick May 1 with no hand-weeding expense and often no tractor cultivations.

Walking in a wheel track with a single-row Earthway seeder.

Seeding

A perfect seeding job is often underrated, but it is still very important to crop profitability. Maximum yields per row-foot are achieved without the unnecessary added expense of thinning. Thinning should be avoided in almost all cases. Having said that, it is not easy to get a perfect stand of plants every time. There are numerous types of seeds; types vary in size or shape, even within the same crop variety; and germination rates for each lot of seed may be different. Careful seeding in the field by hand is very time consuming, and it is unnecessary given the push seeders that are readily available and affordable.

I use two types of push seeders. The first, an inexpensive Earthway Precision, is a lightweight plate seeder that comes with multiple plates for different types of seed. The Earthway is lightweight enough to carry around the farm, has easy-to-switch plates, and does a fine job for many types of seeds. I'll either use a single unit, carefully following the marked lines on the bed, or bolt two or three seeders together, spaced 28 or 14 inches apart, respectively. Perfectly spaced parallel rows are a snap with a multiple-row seeder.

To bolt two Earthway seeders together, drill holes in each marker bar (or use a piece of wood), and use the bars to attach the two seeders together at two points: at the handle and just above the rear wheel. One last bar or strip of wood is then attached to the front kickstands. Wing nuts enable quick changes back and forth from a single-row to a multiple-row seeder. The optional seed plates that Earthway sells are very useful for getting the perfect plant density in the row. I recommend buying a set. I even fine-tune the plastic plates to achieve the correct seed drop by sometimes filing the pickup scoop, or placing tape on every other scoop.

The second push seeder I use is made by Planet Jr. This seeder is much heavier, made of steel with wood handles, not like the lightweight aluminum and plastic Earthway seeder. The Planet Jr. has a seed hopper with a small auger and an adjustable hole in the bottom for the seeds to drop through. There are thirty-nine different hole sizes in all, so fine adjustment is possible. This has obvious advantages when you're trying to achieve the perfect stand of plants. Carrot varieties, for example, all have relatively small seeds, but they can still vary a bit in size. To plant twenty live carrot seeds per row-foot, you can use four or five different hole sizes, depending on the batch of seed.

Two-row Earthway seeder.

The Planet Jr. drops seed down a shoot to either a "single-row" shoe, which can lay the seeds in a narrow band, or to a "scatter" shoe, which spreads the seeds in a 2-inch-wide band. I generally prefer the scatter shoe for all seedings. To achieve a perfect stand every time, I spend as much time as needed finding the correct size seed plate hole for each batch of seed. I bring a blanket or tarp with me to the field to do a couple of test runs before seeding the actual beds. I check the germination percentage, germination date, and past records of the crop to figure which hole size I should begin trying. Then I put some seed in the hopper, open the seeder gate, and push the seeder over the blanket with the rear wheel raised up. Counting the seeds on the blanket gives me a good visual of the seeding. Given the germination percentage, is this the correct amount of seed per foot to drop? I try another pass on the blanket to check or try another hole size and count seeds again. Planet Jr. seeders tend to drop seeds in bunches—unlike the Earthway, which drops seeds in a more even and continuous fashion. With the Planet Jr. there may be a few "clumps" of seed and a few blank spaces, but overall it will be evenly dispersed. Count the seeds over 2 or 3 feet and take an average.

If you've forgotten to bring along a blanket or tarp, an alternative method for seeder

Seeding over a blanket with a Planet Jr. for calibration.

calibration is to open the seeder gate, hold one hand under the seed shoe, and turn the front wheel exactly one-half revolution. One-half of the wheel revolution equals 2 row-feet of travel. Count the seeds that fell into your hand, and adjust the hole size as necessary. The pushing-the-seeder-over-the-blanket method is preferred: It offers a better visual check for seed placement and a longer run for a better average. Take time to adjust the seeder until you are satisfied with the results. The payoff is worth the extra effort: You achieve the highest possible yield without taking the time to thin the stand.

When seeding beds with either type of push seeder, I try not to walk on the bed until I have to seed the middle row. Walk in the space between the beds to seed the outer two rows. Besides the added compaction from walking on the bed, the pressed soil from footprints creates ideal conditions for weed germination. No need to give the weeds any extra help.

When I'm finished seeding, I pour any unused seed back into its bag and write on the bag the Earthway seed plate name or Planet Jr. hole number, plus the date and field location.

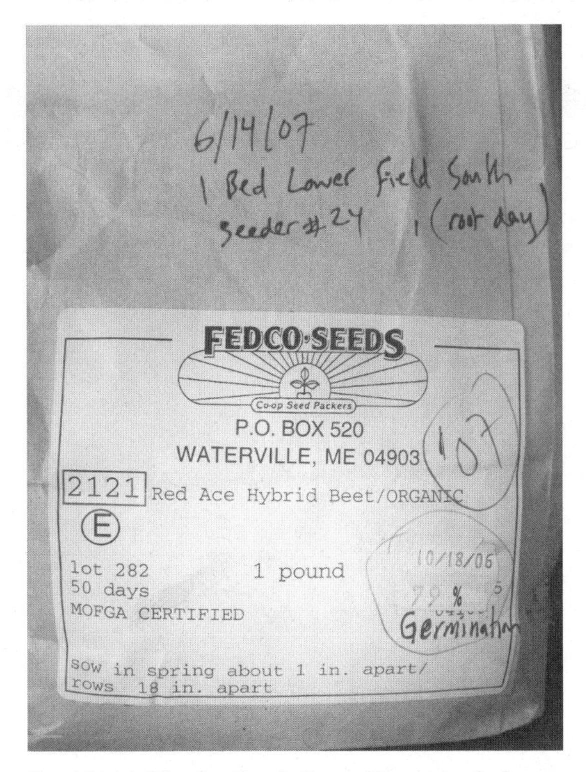

Seed bag with planting information recorded, with germination percentage highlighted.

This is invaluable information to compare future plantings of the same crop. I also record this information into the Crop Journal, along with the amount of bed feet planted, the time it took to seed it, the estimated amount of seed used, and the variety, seed company, and date of seed.

Transplanting

Most transplanting on my farm is done by hand. I do have an older two-row mechanical transplanter, but I use it only for larger blocks of transplanting. Although counterintuitive to most people, transplanting machines do a surprisingly good job of putting tender young plants in the ground. But I have found that hand transplanting is more efficient for areas of ¼ acre or less, given the extra time that is required to hook up the transplanter to the tractor, plus machinery costs, and the fact that one person is dedicated to the tractor seat while two other people actually transplant on the machine.

Transplanting nine hundred plants per hour per person is possible when you're working by hand, depending on the crop, its spacing, and the type of trays the seedlings were grown in. A tractor-mounted transplanter is capable of setting three thousand plants per hour using three people, but it requires an extra hour of setup and disconnect time per use, plus of course the cost of the machinery. Depending on a number of factors (length of beds, spacing, tractor costs, and more), roughly six thousand plants would be the minimum number to justify using a tractor and a transplanting machine.

The key to hand transplanting is *speed.* With small plants, one member of the transplanting crew walks alongside the bed dropping plants on the marked rows, trying to space them within the row. The other person actually transplanting pokes two fingers in the ground with one hand, uses the other hand to push the rootball of the plant into the poked hole, and then moves immediately to the next plant. Boom, boom, boom . . . every two seconds a plant goes in the ground. Thirty plants per minute, or eighteen hundred per hour for two people transplanting. That averages to nine hundred plants per hour per person for the crew of two. No need to fuss and caress and tuck each plant in. Don't worry, plants want to grow! They are genetically programmed to grow and reproduce, like most of nature.

Avoid transplanting on the morning of a sunny day. Plants will do better if transplanted in the afternoon or on a cloudy or rainy day. Weather cooperation helps a lot. Water the plants in if it doesn't rain soon. Take care to plant the crops directly into the parallel lines marked on the bed; all future mechanical weed control depends on it.

Deer Fencing

Deer pressure is so intense on my farm that one night when I left the greenhouse door open, the deer walked right inside and ate off the benches! I'm glad I have a deer-proof latch on the walk-in cooler door.

I can't fault deer for wandering into a field of tasty carrots and sampling a few. Or even more than a few. I'm partly at fault for training generations of deer to become dedicated organic consumers. But I need to make a living, and I don't like killing the deer that happen to stray into fodder heaven. So I try to train the deer to keep away from the crops, and to do it cost-effectively.

I use two-strand poly and metal electric fence wire, 6-foot fiberglass posts, and a good fence charger with an excellent ground. The fence is

Two-strand electric fence sloping away from crop to keep deer out.

portable and quick to move around. It should pack a punch when energized. I place fence posts every 25 feet around the perimeter of the crop I want to protect, inserting the posts at an outward-sloping 30-degree angle. Deer have a hard time with depth perception and are creatures of habit. While a deer can side-jump an 8-foot fence without breaking stride, a slanted fence (or one with a separate outrigger fence line) is a different story. I make a slanted fence with the top wire 4 feet high and the bottom wire 2 feet off the ground. When the deer approach the fence, they get zapped by the top wire and instinctively try to go *under* it. The lower wire shocks them again, and the deer back away.

The 3-D aspect of the slanted fence prevents them from hopping over the top wire, which is only 4 feet off the ground. When I first put a fence in a new spot, I want to make sure the deer are aware of it. I hang a T-shirt that I wore that day on a wooden cross as a kind of scarecrow. The sight and smell of the hanging T-shirt puts deer on the alert as they approach a fence that normally is not there. Without the T-shirt, a deer may bumble through the fence unwittingly. Once trained, though, the deer sense the electric fence and don't try to get inside.

Harvesting

Put unglamorously, farming is a business of "materials handling." Farmers move lots of stuff around, from fertility amendments to cover crop seeds to the bounty of harvest. Tons and tons of materials are moved from one place to another on a farm.

Look at materials flow on your farm. Is it set up to move smoothly and efficiently? A few steps saved here and a few there add up to a lot of steps over twenty years. What improvements to your operation will save you time and effort in the long run? Any tool or technique that increases efficiency pays off from now forward.

A large percentage of materials handling on a farm involves harvesting crops. Lightweight crops like kale aren't as much of an issue as heavier crops like roots. For harvesting root crops, I use two harvest aids: a bedlifter and a chain digger. The bedlifter is a simple implement with no moving parts that loosens the soil in a bed of root crops such as carrots or parsnips. Plants are pulled easily from the ground, with no shovel necessary. An angled blade travels under the crop and fractures the soil by moving it slightly upward. The bedlifter also works to remove plastic mulch.

Bedlifter to loosen soil in beds.

My chain digger is a modified potato digger, widened to lift an entire bed at once. The nose plate of the digger travels under the crop, and a conveyor chain lifts soil and the crop out of the ground. Soil falls through the chain links, and the crop falls onto a soft bed of soil for easy pickup. I use this digger for lots of root crops, perennials, and its intended use, potatoes. Many hours are saved when picking up

Mowing echinacea.

a crop that is on top of the soil and in plain sight. Unmarketable produce can be left in the field without the labor of first pulling it from the ground.

Chain-dug root crops are picked into bushel baskets or poly-weave grain sacks and trucked back to the pack house. Extra hands are hired during these busy times. This enables us to get a lot accomplished during the growing season so we can store and sell produce in the off-season, spreading out our sales throughout the year.

Bags or bushels are unloaded from the truck to the storage area using a bucket brigade. A few crew members pass produce from one to another in a chain instead of each person walking from the truck to the storage area and back for each bag or bushel. Fewer steps are walked, and it makes harvesting a more social activity.

With stackable containers like boxes, I always use a hand truck. Once boxes are stacked on the floor, no more lifting is necessary. Simply tip the stack forward slightly, insert the hand truck, and move the containers. No need to bend over and pick up the same boxes over and over again.

The construction slogan *Never walk across a job site empty-handed* definitely applies to

Digging roots with a chain digger.

farming. Crew members should always remember to bring picking containers, knives, and rubber bands with them into the field. It seems obvious enough, but I wouldn't mention it for no reason. As with any farm work, I bundle tasks for efficiency, whether it's stapling up boxes to pick into or picking all the parsley for orders and then moving on to the next crop on the list. When picking bunched crops, I use #30 rubber bands (except for broccoli). I count out ten bands and place them on my opposite hand's pinkie and ring fingers. If I need to pick eighty-five bunches of cilantro for orders, I make ten bunches and leave them in a pile of ten. Ten more rubber bands go on my fingers and I keep picking. I can easily see the piles of ten bunches behind me to know when I have reached my target of eighty-five. I do a recount as I place the bunches in the box to go to the pack house.

In the pack house, again look at materials flow. Does the produce move smoothly from the truck to the washing/sorting area, and then to storage? Are materials moving in one direction or back and forth? Again, a few steps saved adds up over many years of farming.

Besides the flow, look at how you grade your produce. Vegetable washers do a great

Barrel washer for quick and thorough cleaning of root crops.

job speeding up the process of cleaning produce, but crops still need to be graded once they're clean. I find some crops with multiple grades harder to pack out and more prone to error. For example, we grade carrots into A's, A minuses, B's, juicers, horse feed, and compost. When sorting from a pile of washed carrots, I find it more efficient to pull one type at a time for a while, then switch to another grade. Otherwise, you have to make a judgment call every time, and misgrading is more likely.

Greenhouse Efficiencies

Most farms that raise crops from transplants produce their own seedlings in a greenhouse. Production systems vary widely. Look at your greenhouse and notice how materials flow. I find that the greatest efficiencies come from moving potting soil, moving plants, and watering plants.

Potting soil is best stored near its next use: filling plant trays or making soil blocks. A few steps saved here every day saves a lot of hauling. I installed a trapdoor in the greenhouse

Overhead sprinklers in middle and sides of greenhouse. Note trolley rail hung by C brackets.

roof so potting soil can be easily loaded directly onto the potting bench inside. The potting soil is dumped outside the greenhouse next to the trapdoor to minimize the distance it has to be moved.

Once plants are started, they are moved with a simple trolley cart hanging from an overhead rail. I love my trolley. Plants can be moved within the greenhouse, outside, or to another greenhouse. Trolleys are very effective, low-maintenance, and inexpensive. Before the trolley, greenhouse workers would carry one flat in each hand hundreds of feet at a time. With the trolley, twenty-five flats can be easily moved with a gentle push. Seeded flats are moved from the potting bench to growing benches and then, when ready, moved outside or to another greenhouse. Loading the van for deliveries is a snap. I go "shopping" up and down the greenhouse aisles, filling my trolley with the plants I need and then moving them outside to a waiting van.

As with moving potting mix and plants around, watering seedlings should also be as efficient as possible. When I first started farming, I thought watering was a job anyone could

HEARING LOSS

Farmers often work around machinery and power tools, sometimes only in short increments of time. Two cuts with a circular saw or half an hour on a tractor doesn't seem like much, but the noise from these tools is no less intense. Loud noises can cause hearing damage. Damage to hearing does not heal. Once you have lost some of your hearing, it won't come back. Hearing protection is mandatory for everyone on our farm who is exposed to loud sounds (as well as eye protection when needed). I bought earmuffs for each tractor, the lawn mower, and the hand tiller, and I leave them with these machines. The earmuffs become part of the machine's use. Three more earmuffs live in the shop for use with power tools. Don't learn the hard way like I did and lose some of your hearing. Get some ear protection and use it every time.

do. Not so. Watering is an important organic cultural practice that keeps plants healthy and strong. Sustained attention and knowledge are needed so as not to over- or underwater plants. That said, watching water flow out of the end of a hose every time plants are thirsty may not be necessary or the best use of your time. Overhead sprinklers controlled by a water timer can do an adequate job for 75 percent of your watering needs. More precise hand-watering makes up the other 25 percent. Sprinklers come in different shapes and sizes, and their selection depends on your greenhouse setup. Talk to your greenhouse supplier and other farmers before deciding on which sprinklers to get.

A recent innovation in our greenhouses is a system of bottom-watering benches. These benches are sloped slightly and covered with metal troughs that the plant trays nest into. Water is pumped from a stock tank to the top of the bench, where it flows underneath the trays of seedlings. The soil soaks up the water by capillary action through the bottom of the pots, just like a dry sponge on a wet countertop. Excess water flows past the trays and is collected in a gutter that empties back into the stock tank.

Another type of watering table is the ebb-and-flow system. Instead of water *flowing* under

THE EDGE EFFECT

Efficient use of your land means maximizing the use of your fields. And this sometimes means maximizing the field itself. Weeds often encroach on the outer edges, shrinking the field's usable area. It may not seem like a lot, but this outermost edge adds up to a lot of land. Think of the old-style LP vinyl records, where the outermost groove is so much longer than the innermost. One or two extra tractor passes around the edge of a field dramatically increases the area of your field. By reducing your field margins, you may postpone expansion to another plot of land altogether.

the plants, the bench top is like a large shallow tray and is set perfectly level. Water fills the tray to flood the plants. After all the seedlings are saturated with water, the bench is drained completely. Ebb-and-flow systems are a little more expensive to purchase than flow-through bottom-watering tables. Each type will need adjustable leg supports on a stable base. Both systems have some drawbacks. Recycling water can spread disease if disease has gotten a foothold in your greenhouse. And plants on each bench should be of similar size for efficient watering, which requires more labor moving plants around from bench to bench. Depending on the amount of plants you produce, sprinklers may be the most cost-efficient labor saver.

– 10 –
Write Your Own Business Plan

My dog just gave me a look that said, *Richard, why do you choose to write about the driest subjects? Wouldn't you rather play fetch?* I dodged the first question by responding that only humans would understand the answer.

Many farmers are often turned off and intimidated by the thought of writing a business plan. After all, business plans are the domain of "real" businesses, not farms. Farms are farms, where plants and animals are raised and tended. Farmers work the soil and cultivate living things. Farms are also our homes. There is frequently a disconnect between farms and business plans. Business plans are documents for businesspeople in offices, full of pages with text and forms full of numbers; they're lifeless dust collectors.

Business plans often have a dry and dusty image because most businesses lack a feeling of "ownership" with the plan, especially if an outside person has authored it. Yet in its best form, a business plan is actually very alive: It is a thoughtful representation of you, your farm, and your future plans. Think of a business plan as a *road map* of where you want to go. Your destination is somewhere off in the future with unknown turns. You begin plotting your course from where you are right now, and you consider various routes to determine the best one to reach your destination. A business plan can be just that simple.

Most importantly, a farm business plan is designed for *you*. The thinking that goes into a business plan and the information represented there will help you get where you want to go. It takes time to reflect and to direct your farm into the future, and the thoughts you arrive at are valuable ones you don't want to forget. Writing your thoughts and ideas down on paper makes them more real and easier to remember when you're reviewing them at a later date. The very act of *writing* a business plan validates its reality.

You may have noticed earlier in this book my oft-repeated refrain: *Farming is a business, and the business aspect of farming is often neglected.* A business involves income, expenses, and profits, all necessarily expressed in numbers. Besides using words, business plans speak in the universal language of figures, often in three common forms: a Profit and Loss Statement, a Balance Sheet, and a Cash Flow Projection. These financial forms will be explained later on in this chapter. I'll start with the written-word topics: descriptions, analyses, and planning.

Business plan writing can be a lengthy or short exercise and may involve businesses that are small or large, farm-based or not. A manifesto it need not be. A complete plan might be only five to ten pages from start to finish, including financial forms. Plans can certainly be longer if needed. Again, the business plan is *your* road map. The more involved you are with its creation, the more ownership you will feel with the plan. No one else needs to see your business plan, but occasionally it may be appropriate to show others this snapshot of (and vision for) your business. Examples of people who might need to see your plan are a banker you want to borrow money from, a potential business partner or investor, or a family member. You might also bring in a friend or hire someone as a coach to jump-start or help develop the business planning process, which is a great idea. Just remember to ask him or her to maintain a level of confidentiality.

The Skeleton

My primary goal in this chapter is to demystify the business planning process and show you a skeletal sketch of all the business plan components, so that you are more inclined to write a plan for yourself. Many written resources are available to further your efforts in crafting a business plan, as is personal help from various business advocacy groups. Check these resources if you want to delve deeper into the subject.

Here's a business plan skeleton to give you a quick overview.*

1. Cover page with farm name, contact information, and date.
2. Table of contents.
3. Executive summary: This brief summary of your plan is written at the very end of the process.
4. Descriptions:
    - *Basic farm description*: Location, size, ownership type, who is involved, a short history of the farm, a list of buildings and fields, a farm map, geographic relation to other towns or markets.

*Modeled after the Vermont Farm Viability Enhancement Program business planning guidelines. For more information, visit www.vhcb.org.

- *Farm products description*: What is sold, to whom, and amounts. A Marketing Chart from chapter 2 or a Sales Spreadsheet from chapter 3 says it all in numbers on one page. Also included is a short history of farm production and current promotional efforts.

5. Analyses:
   - *SWOT analysis*: *SWOT* stands for Strengths, Weaknesses, Opportunities, and Threats. Brainstorm a list for each category.
   - *Management analysis*: Who manages what on the farm. This analysis addresses aspects of farm production, finances, employees, and marketing.
   - *Market analysis*: The current status of all sales venues and potential trends.
   - *Enterprise analysis*: Budgets for your biggest sales items (or all of them) to determine profitability.
6. Planning:
   - *Proposed ideas and strategies*: Given your SWOT analysis, what direction will your farm take?
   - *Implementation of ideas and strategies*: A timeline for different projects to undertake, including the who, when, how, and why of each planning change. Budgets for changes show cost/benefit impact.
7. Financial forms:
   - *Profit and Loss:* A budget for the whole farm, showing net profit for the year.
   - *Balance Sheet:* A statement of your net worth—what you own minus what you owe.
   - *Cash Flow Projection:* Your road map in numbers. What should all your income and expenses look like in the next five years?

That's it! No rocket science in there.

Some Details

Begin the business planning process by writing the descriptions. Then tackle the analyses, planning, and financials. Even though the summary is placed in the leadoff position, just after the table of contents, it is written only after all the other parts of the business plan are complete, much like the introduction to a book. So save the summary for last. Below is an elaboration of the business plan skeleton, adding a little substance to the bones.

1. **Cover page:** Your farm name in bold letters, with farm address, phone, website, e-mail address, and date. Cite the authors of the business plan.
2. **Table of contents:** This can follow the preceding outline, or you can make

your own. It can be written now or after the plan starts taking shape.

3. **Executive summary:** A short encapsulation of what the business plan is about. What is its focus? What important issues face the farm? What changes are proposed? I recommend including in the summary a farm mission statement that puts the big picture front and center. Financial summaries of net worth, gross sales, and net farm income for the current or past year are clear indicators of where the farm is financially at the time of writing.
4. **Descriptions:** The basic farm description can be one to three paragraphs long, often expressed in the words that are right on the tip of your tongue. You know your farm best: How would you describe it to *someone else*? Elaborate on any part of the farm that you feel is important, but be sure to include the points I mentioned earlier. The farm products description can be depicted in your one-page Marketing Chart from chapter 2 or the Sales Spreadsheet from chapter 3. These tables say it all, but a narrative of what you sell and to which markets is helpful to a reader other than yourself. A short history of when you started, at what scale, and how production has changed over the years illuminates trends. Include current marketing efforts (flyers, ads, website) as a record of what you are doing now for promotion.
5. **Analyses:** Start by brainstorming ideas for your SWOT analysis (Strengths, Weaknesses, Opportunities, and Threats). Ask members of your farm's inner circle to contribute ideas. Write down all their thoughts, remembering that there are no right or wrong contributions when brainstorming. Don't worry about being repetitive or out of order. The list can be refined later on. Here are some examples with a few possibilities:
   - *Strengths:* Location near a good market, fertile soils, energetic management.
   - *Weaknesses:* Lack of irrigation, lack of labor, rocky ground.
   - *Opportunities:* New organic retailer opening in town, land lease becoming available, localvore movement thriving.
   - *Threats:* Competition in the marketplace, higher energy costs, new regulations.

   Look at your long list of SWOT items and organize them. These items will be the foundation of an action plan for your farm. Look at each category and try to condense all the items into one sentence. Do this for each category.

Though concise, this narrative says a lot about your farm business.

- *Management analysis:* Evaluate the human resources that run the farm business. Who does what in managing farm production, finances, employees, and planning? Are skills sufficient or do some areas need attention? If professional development training would help, specify what types and for which people.
- *Market analysis:* Look at the current status of each marketing venue and ask what will likely happen to it in the future. Is demand for product steady, gaining, or shrinking? What are the prospects for sales prices? Assess your competition; will it affect your business? How will your product be distributed and promoted? Spur your thoughts by referring to the Marketing Circle in chapter 5.
- *Enterprise analysis:* Paramount to any business decision is how an idea pencils out in numbers. Enterprise analysis should begin with your top sales items to make sure they are profitable. Take the time to do a few simple budgets to see where the money is currently flowing. Chapters 3 and 4 give examples of budgets. Use these to help you analyze your farm products.

6. **Planning:**
   - *Proposed ideas and strategies.* As the namesake of the whole document, this section is the road map to possible changes to your farm business. What ideas and plans surfaced when you completed your SWOT analysis? What strategies do you want to try? This section encompasses the heart of your business plan and its overarching purpose: What is the focus of your plan for the future? Review your strengths and opportunities and use them to your advantage. Address issues identified as weaknesses and threats, and look at possible countermeasures.
   - *Implementation of ideas and strategies:* Chapter 6, "Effective Management," described assigning each important task a block of time on a calendar to make sure it gets completed. Implementing your proposed business plan ideas is no different. When will the task be tackled? Who will do it? How long will it take, how much money will be needed, and where will that money come from? How will the proposed change affect overall farm profitability? Sketch out some simple financial calculations to back up

your planning ideas. An irrigation system may cost $4,000, but over a ten-year life span it costs only $400 per year (not including interest or any repairs). Does the prospect of increased yields, more sales, or fewer other irrigation expenses justify the annual $400 for a new irrigation system? Contemplate the financial aspects of your decisions and write them down.

Financial Statements

The next part of the plan covers three types of financial statements that speak clearly in the language of numbers.

Profit and Loss Statement

This is also called an Income Statement. In its simplest form, a Profit and Loss Statement lists all types of income and expenses for the farm and the resulting net profit (or loss). Profit and Loss Statements follow the same format as the Chart of Accounts as seen in chapter 7, with sales and income venues listed first, followed by expense categories. Profit and Loss Statements always cover a *specified time period.* Often the time period is the calendar year of January 1 through December 31, but other time frames are possible, like quarterly or monthly statements used for closer monitoring of finances. Table 10-1 is a sample of a simple annual Profit and Loss Statement.

TABLE 10-1: Profit and Loss Statement January 1, 2008–December 31, 2008

Income	
Sales: Wholesale	16,500
CSA	12,000
Farmers' market	14,200
Total Sales	**42,700**
Expenses	
Advertising	120
Donations	50
Fertilizer	960
Fuel and oil	1,280
Greenhouse supplies	2,340
Insurance: farm share	810
Interest expense: farm share	1,800
Internet, website	360
Livestock expenses	750
Miscellaneous	3,010
Office	440
Payroll	10,200
Professional services	560
Rent paid	500
Repairs and maintenance	4,070
Seed, plants purchased	1,270
Supplies	1,320
Taxes: farm share	1,230
Trucking, freight	540
Utilities: farm share	880
Total Expenses	**32,490**
Net Profit	**10,210**

CASH AND ACCRUAL ACCOUNTING

With financial accounting, two basic methods are used: cash and accrual. In explaining the two methods, I'll use the calendar year as the time frame. Think of cash accounting as the way your farm checkbook register works. If you paid for something during the calendar year, you would subtract the check amount from your checkbook balance and record it for that year. If you were paid for sales and deposited the money in the bank during that calendar year, you would record that as income for the year. Many farms use this type of cash accounting. But what about sales you make in December but don't get paid for until the following January? Or expenses you incurred in December but don't receive a bill for until after the New Year? With cash accounting, the bank deposits and checks you write in January will be recorded in January's year, not the year the transaction actually occurred. Cash accounting is simpler and more familiar because it is usually the way we manage our personal checkbooks.

With accrual accounting, on the other hand, the transaction dates of sales and expenses matter; they are recorded in the year that the transaction actually happens and use inventory to reflect expenditures. Many farm suppliers offer discounts if farms prepay in December. It is an incentive to get sales. As an example, I may buy $2,000 worth of potting mix in December 2008 for a discounted price. Say this pre-buy offer was a ***onetime deal;*** what effect will it have on my Profit and Loss Statement?

With cash accounting, my net profit in 2008 would be $2,000 less because I paid for the potting mix in December 2008. My 2009 net profit would be $2,000 higher, because I wouldn't need to spend anything on potting soil in 2009 (it was pre-bought in 2008). These two years' Profit and Loss Statements don't reflect what really happened.

Accrual accounting paints a more accurate financial picture for each year. In 2008, a check was written for $2,000 and my inventory (of potting soil) increased by $2,000. Thus, when accounting for inventory, my 2008 net profit would be unchanged by the potting soil purchase.

In 2009, no check was written for potting soil, but accrual accounting's inventory shows $2,000 of soil on January 1 and $0 on December 31. Inventory was reduced by $2,000 (spent) and the net profit for 2009 would reflect the pre-buy potting soil purchase. Both 2008 and 2009 Profit and Loss Statements accurately represent what happened:

- In 2008, a $2,000 check was spent on soil inventory.
- In 2009, $2,000 of inventory was spent but no check was written.

Anyone ready to play fetch with the dog now?

This is a cursory overview of accounting principles. Whichever system you decide to use, be consistent so you can compare apples to apples.

Balance Sheet

A Balance Sheet lists what you own (assets), what you owe (liabilities), and what is left over (net worth). Unlike a Profit and Loss Statement, which covers a span of time like a calendar year, a Balance Sheet measures assets and liabilities *at a specific moment of time*—say, midnight on December 31, or right *now*. The net worth from different years' Balance Sheets can be compared to see if you are gaining or losing financial ground over time.

Balance Sheets can be calculated solely for farm assets and liabilities, but including personal finances such as savings accounts is more comprehensive and provides a much clearer financial picture. Inventory is included when measuring assets, whether personal finances are included or not. Farm machinery, greenhouses, and farm supplies all have value and are listed with fair market prices, not original purchase prices. In the previous example of pre-bought potting mix in 2008, this isolated transaction is portrayed in the two Balance Sheets shown in table 10-2. Notice that the net worth is not affected from 2008 to 2009 by the potting soil purchase. Again, this assumes the pre-buy was a onetime deal in December 2008 and not repeated in December 2009.

The Balance Sheet will often show where farm profits are spent. In chapter 8, I gave an example of a farm with a $36,000 anticipated net profit at year end and showed two possible options: pay taxes on the $36,000 *or* buy a $12,000 tractor before year end and pay taxes on the resulting $24,000 net profit. Table 10-3 shows what simplified Balance Sheets would look like in the two scenarios.

Notice how the net worth *increased* with the tractor purchase while the net profit *decreased*. The tax savings of the reduced net profit shows up in the Balance Sheet. Farm purchases may create a low year-end net profit and mislead someone into thinking that the farm is an unprofitable operation. But the Balance Sheet would reveal an increase in net worth from the farm purchases despite the low year-end profit.

Now for the full monty—a sample Balance Sheet with short-, medium-, and long-term assets and liabilities is shown in table 10-4.

Notes can accompany the Balance Sheet to show the breakdown of individual line items like inventory or supplies. A separate list of farm

TABLE 10-2: Balance Sheets

Assets:	On December 31, 2008	On December 31, 2009
Farm checking balance*	$8,000	$10,000
Inventory potting mix	$2,000	0
Total assets	**$10,000**	**$10,000**
- Liabilities	**0**	**0**
= Net worth	**$10,000**	**$10,000**

*Arbitrary balance for illustration purposes.

TABLE 10-3: Balance Sheets

Assets:	No tractor purchase	Purchase $12,000 tractor
Net profit	$36,000	$24,000
Tractor	0	$12,000
Total assets	$36,000	$36,000
- Liabilities	0	0
Income taxes	$12,000	$8,000
= Net worth	**$24,000**	**$28,000**

TABLE 10-4: Balance Sheet on December 31, 2008

Assets		**Liabilities**	
Current assets		**Current liabilities (due in less than 12 months)**	
Farm checking	2,350	Operating loan balance	0
Savings account	4,000	Line of credit balance	500
Accounts receivable	520	Credit card balance	860
Crop and feed inventory	800	Other	0
Farm supplies on hand	1,250		
Total current assets	8,920	Total current liabilities	1,360
Intermediate assets		**Intermediate liabilities**	
Farm machinery	26,300	Tractor loan balance	8300
Farm vehicles	10,000	Truck loan balance	4500
Livestock	0	Other	0
Retirement accounts	35,000		
Total intermediate assets	71,300	Total intermediate liabilities	12,800
Long-term assets		**Long-term liabilities**	
Farmland	40,000	Long-term loan 1 balance	110,000
Farm buildings	42,000	Long-term loan 2 balance	70,000
Farmhouse	120,000		
Total long-term assets	202,000	Total long-term liabilities	180,000
Total assets	**$282,220**	**Total liabilities**	**$194,160**
Net worth: $88,060			

equipment with each piece's market value totals up to the figure shown on the Balance Sheet. I do not want to take the time necessary to explain each line of the Balance Sheet. I'd rather have you see and digest a sample in complete form and possibly use it as a model.

Many financial ratios can be calculated from numbers on the Balance Sheet to indicate the health of the business. This book's accompanying CD contains a workbook from the Vermont Farm Viability Enhancement Program that automatically determines some indicators once other financial forms are completed. But the simple tracking of your net worth from year to year speaks loudly in financial terms of where you have been and where you are going.

Cash Flow Projection

Just like the planning for profit exercise in chapter 2, a Cash Flow Projection is a road map of numbers. Cash Flow Projections look much like Profit and Loss Statements, but with future years tacked on. Depending on changes you plan to implement, sales of various products will go up or down, as will different types of expenses. These changes will affect your bottom line, or net profit. Doubling production may double sales without doubling expenses. Your Cash Flow Projection will show this increase in net profit.

Projections are educated guesses of the future, and until someone discovers a crystal ball that works, you must plan as best as you can to achieve your goals. Start with your last-known sales and expense figures as a benchmark. How will planned changes from section 6 of your business plan affect each sales and expense item? Say the plan advocated the purchase of a cultivator, flame weeder, and barrel washer to increase production efficiency. This added cost in 2009 will hopefully enable more efficient use of labor. More crops can then be produced with the same number of employees. The more efficient production from the new tools, coupled with other efficiencies mentioned in this book, allows increased sales to be generated with the same or decreasing expenses.

The term *cash flow* can also refer to how much money is available at any given time during the course of a year, month, or week. As opposed to a salaried worker with a regular weekly paycheck, farmers have notoriously "lumpy" cash flows. Think of a lumpy cash flow as a checkbook balance that resembles a roller-coaster track over the course of the year. Farms incur many expenses early in the year in order to produce lots of income months later.

Uneven cash flow is a concern because bills have to be paid regardless of how little money is in the farm checkbook. Try hiring employees and informing them that their paycheck will be six months late. That's a hard sell. Cash flow shortfalls can be overcome by borrowing money via short-term loans or a line of credit from the bank. You pay for this service of borrowing money in the form of interest payments. Ideally, your farm will generate profits that can be salted away to lend back to yourself when needed, but until then, some debt may be incurred.

A very important point needs to be made regarding cash flow: *A profitable farm operation*

TABLE 10-5: Cash Flow Projection 2009–2012

	2008	2009	2010	2011	2012
Income					
Sales: wholesale	16,500	22,000	26,000	30,500	34,000
CSA	12,000	14,000	16,000	18,000	20,000
Farmers' market	14,200	16,000	18,000	20,000	22,000
Total sales	42,700	52,000	60,000	68,500	76,000
Expenses					
Advertising	150	150	150	150	150
Capital purchases	0	4300	0	0	0
Donations	50	0	0	0	0
Fertilizer	960	1,500	1,900	2,400	3,000
Fuel and oil	1,280	1,600	2,000	2,500	3,100
Greenhouse supplies	2,340	2,200	2,400	2,500	2,600
Insurance: farm share	810	850	875	900	925
Interest expense: farm share	1,800	1,600	1,400	1,200	1,000
Internet, website	360	360	360	390	390
Livestock expenses	750	800	850	900	950
Miscellaneous	3,010	3,000	3,000	3,200	3,200
Office	440	480	500	520	540
Payroll	10,200	9,000	9,500	10,000	10,500
Professional services	560	560	580	600	620
Rent paid	500	500	500	500	500
Repairs and maintenance	4070	4,000	4,000	4,500	4,500
Seed, plants purchased	1,270	1,400	1,600	1,800	2,000
Supplies	1,320	1,500	1,700	1,900	2,100
Taxes: farm share	1,230	1,250	1,250	1,300	1,300
Trucking, freight	540	700	800	900	1,000
Utilities: farm share	880	900	920	940	960
Total expenses	32,490	36,650	34,285	37,100	39,335
Net profit	**10,210**	**15,350**	**25,715**	**31,400**	**36,665**

can be put out of business by a lack of cash flow. On the surface, this doesn't seem to make any sense. If a farm is profitable, why would it be forced out of business? But think of what causes businesses to fail: They can't pay their bills. The scenario goes like this: No money to buy supplies, payroll can't be paid, late loan payments trigger default action, property gets sold, and the IRS jumps to the front of the creditors' line to collect any taxes due. Not a pretty picture. A profitable business would be able to make all the payments in the long run, but without interim cash flow it can be forced to shut down. Without loans or cash reserves to tide the farm over until sales come in, your creditors can claim what's theirs and start the legal process to get it back. Remember this when you think of cash flow and line up necessary funds *before* you need them.

Write the Summary

Now all the pieces are in place for you to write the summary that will lead off your business plan. Describe the purpose of the plan succinctly. What are some crucial issues facing the farm, and what are some proposed changes to the operation? Include key financial data like current sales, net income, and net worth. How might these financial indicators change over time by implementing proposed ideas and strategies? I recommend including a mission or goal statement for your farm business as part of the summary. Refer to your work from the exercises in chapter 1 to put your "big picture" up front in the business plan.

Voilà! Your business plan is complete. You now have a road map for the future of your farm. Congratulations. In the process of writing the plan, you've likely already set many changes into motion. Make your business plan vital and relevant by referring to it at least semiannually or annually. Don't park it on the shelf as a static, finished product. Use it for future planning and incorporate any new ideas and strategies. Planning is an ongoing process. Enjoy the ride.

– 11 –

Planning for the Inevitable: The Ultimate Conclusion

When asked about the plans he'd made for his death, the elderly man quipped, "Death? I plan on being immortal. It's worked so far!"

A book on farm business could easily avoid the topic of estate planning. What does death have to do with farming? Conventional wisdom maintains that only old people need to think about end-of-life matters. I used to think that such a topic would be taught by your parents or by books specializing on this matter of grave concern. But like the material in this book's preceding ten chapters, I had to learn the lessons along the way. None of it was programmed into my genetic code. I want to shed light on this dark subject because I personally feel it is a topic too important to avoid.

The benefits of estate planning are not for you; they are for the people you care most about, those who remain to shoulder the burden of your death. We all know we will die someday, but we still put off dealing with it. Death is not a matter of *if,* but rather *when.* I am reminded of my own mortality whenever a friend or family member dies. Like most people, after I experience the loss of someone close to me I temporarily reflect on the meaning of life and death, but then slowly return to my normal life and ways of thinking.

The way this topic fades from the forefront of my thoughts is the reason why I could so easily procrastinate about planning my own end. My motivation to write a will came when my wife and I took a trip without our kids. What would happen to them if we were both in a plane crash and died? Far from being ghoulish, we felt that the only responsible thing was to plan for such an event.

The purpose of estate planning is to consciously direct your belongings the way you want them to be distributed upon your death, not necessarily the way the government wants to. If you die without a will (or other will substitute), the laws and court system decide what will happen with your assets, with possibly a larger tax bite taken out. With proper estate planning, your designated beneficiaries receive what you want them to inherit, and you avoid overpaying taxes. A written will can make this a reality.

Wills are legal instruments. You'll probably want to spend a few hundred dollars and hire a lawyer to draft the document in legalese. Educate yourself first, however. The book *Estate Planning for Dummies* by N. Brian Caverly and Jordan S. Simon covers the subject well in plain English. You'll find that many options exist for estate planning, ranging from simple to complex, but don't let the many alternatives deter you from taking action.

The document used in estate planning is usually a will or a trust. There are many types of each, as well as other options. Setting up a trust is often more complex than writing a will, but trusts do have their benefits. I'll keep the discussion very brief and talk about the simple will. I am not an attorney, though, so don't take the following as ironclad legal advice.

Three separate documents are usually drafted when making a will, described below.

- **The will:** Generally this document applies only to *you* and deals with how your assets will be distributed, and to whom. The named beneficiaries can be people and/or institutions. You'll need to appoint an executor (and an alternate)—someone in charge of executing your will after your death. Possible candidates are your spouse, a family member, a good friend, or your lawyer. Executors need to deal with legal responsibilities and perform actions at an often emotionally charged time. Make sure your executor knows this. Your will also spells out directions for care of any dependent children.
- **Durable power of attorney:** This is a document where *you* (as opposed to the court system) appoint someone else to act on your behalf for general or business matters if you become incompetent or incapacitated. For instance, if you are in a coma, your appointed agent can carry out your day-to-day affairs, such as paying your bills. The agent doesn't necessarily have the power to decide your health care directives; this is described next.
- **Health care directive:** In this document, you appoint someone to make decisions regarding your health care decisions if you become unable

to do so yourself. It may or may not be the same person that you have designated as having durable power of attorney. You can also state your wishes that certain types of care shall be provided (or not provided) if you are terminally ill or about to die.

Once the documents of the will are signed, they need to be placed where someone will find them after you die. Your lawyer customarily retains a copy. A good friend or relative may also keep a copy or just know where to look for it. My wife and I keep our will papers in a metal box inside the chest freezer in our home's basement. Our insurance agent informed us that a freezer in the basement is one of the last places destroyed by fire and is not an uncommon place to store valuable documents. Our adult children and some select friends know of this location, and, in case they forget, the metal box is clearly labeled for eventual detection.

I have also outlined an obituary for myself. Constructing time lines for your life is a lot easier for you than anyone else, and preparing this outline yourself means the element of grieving doesn't cloud the task. There's no need to have a ready-to-print obit; just frame the main points in your life. Include such things as place and year of birth, where you lived, schooling, vocations, important dates and events, significant others, your interests, family lineage, a good photo of yourself, or anything else you feel is important. What an exercise it is to reduce your life so it can fit into 12 column-inches in one edition of the local newspaper. Whew! It makes you appreciate living in the moment. Remember that the obituary outline's purpose is to make necessary tasks a little easier for those who have to deal with your passing.

Last, think about what you would like to have done with your remains. Do you have a final resting spot in mind? Do you need to take actions now if you want a cemetery plot? Do you want a headstone or marker? Is a funeral home going to be involved? Ponder these questions and write down your thoughts. I keep this information and the obituary outline in the metal box with my will documents.

A Parting Thought

I was introduced at a conference a few years ago as Richard Wiswall, loan officer for the NOFA VT* Revolving Loan Fund. Never would have I imagined that the two words *loan officer* would become a suffix for my name. And here I am now, writing a book on the business of farming, as I risk becoming known as an UBCOF: an Über Bean Counting Organic Farmer.

In my defense, though, bean counting in its highest form can offer farmers a means to achieving their financial (and personal) goals. For years I struggled to make a living at farming until I realized that economic security was an important need and goal of mine. Various forms of bean counting were needed to achieve this goal.

*Northeast Organic Farming Association of Vermont

While the focus of this book has been on how to make an organic farm profitable, I want to stress that the pursuit of profits is only part of life's overall picture. Yes, money provides economic security, but don't let it overshadow what's really important in life. In the end I find that family, friends, and the pursuit of a meaningful life (whatever that means to you) always trump making a healthy profit.

Finally, I urge you to go back to chapter 1 and revisit the goal-setting exercises discussed there. Look at your goals. Brainstorm ideas and envision long-range plans. Make a master to-do list, prioritize the tasks, block some time out to do them on a calendar, and follow through to make sure they get done. Be the farmer you want to be.

Appendix

Vegetable Farm Crop Enterprise Budgets

Workbook Index

Workbook Index

Worksheet 1	Labor, Delivery, Farmers' Market, and Overhead Costs
Worksheet 2	Greenhouse Flat Costs
Worksheet 3	Greenhouse Costs: Bedding Plants and In-ground Tomatoes
Worksheet 4	Tractor, Implement, and Irrigation Costs

Crop Budgets:	Net Profit per 1/10 Acre	Extrapolated to Net Profit/Acre
Basil: bunches	$3,560	$35,603
Beans: bush	-272	-2,720
Beets: roots	825	8,253
Broccoli	116	1,157
Cabbage	581	5,806
Carrots: roots	1,405	14,046
Celeriac	1,366	13,659
Cilantro: bunches	1,656	16,561
Corn: sweet	-192	-1,922
Cucumbers	153	1,531
Dill: bunches	1,623	16,232
Kale: bunches	2,463	24,630
Lettuce: heads	791	7,905
Onions	611	6,110
Parsley: bunches	4,742	47,425
Parsnips	1,384	13,844
Peas: snap	-217	-2,165
Peppers: bell	1,556	15,556
Potatoes	261	2,610
Spinach	1,015	10,147
Squash: summer	787	7,867
Squash: winter	87	869
Tomatoes: field	1,872	18,724
Tomatoes: greenhouse		Not applicable

Notes on Net Profit/Acre: Refer to chapter 4 for more information. I've tried to make all crops comparable and as accurate as possible. The budget numbers represent very efficient crop-production techniques, and so numbers may be on the high side for some net profits. Crops grown in smaller blocks and/or raised less efficiently will lower potential net profits.
All field-crop budgets include one watering by an irrigation system. Net profits/acre are extrapolated from the 1/10-acre profits that are figured in the crop budgets. When entering different sales prices or yields in the budgets, net profit/acre will be affected dramatically. All these budgets figure some of the costs from a hypothetical 5-acre farm with two greenhouses selling crops both wholesale and retail. The budgets include approximate costs for marketing, delivering, and overhead. Depending on the crop, some are budgeted being transplanted in plastic mulch, some direct-seeded, some with row covers, and some not. Your numbers are the best numbers. Copy the budget sheets and enter your own data to find out where your profit centers are.

Worksheet 1

Worksheet 1

Labor, Delivery, Farmers' Market, and Overhead Costs to Use in Calculating Crop Budgets

Labor Costs:

	Manager	Crew	Composite crew 1:3
Average hourly rate:	10.00	10.00	10.00
Employee taxes: 7.51%	0.75	0.75	0.75
Workers' comp: 8%	0.80	0.80	0.80
Nonassigned time: 10%	1.00	1.00	1.00
SEP-IRA: 25%			0.00
Labor costs/hour:	**12.55**	**12.55**	**12.55**

Labor costs are critical to calculating crop budgets. The farm's labor cost per hour is more than the employee's wage when employer taxes, workers' comp insurance, and nonproduction time (meetings, cleanup, maintenance) are added in. The SEP-IRA is an optional retirement plan, which is an added cost for certain qualifying employees (see chapter 6). If a farm manager is at a different pay rate, a composite rate per hour can be used. This worksheet assumes a ratio of 3 crew workers to 1 manager. For simplicity, all labor is paid the same rate in these crop budgets.

Delivery Costs:

	Produce	
Labor: load truck(s) and travel	25.10	@12.55/hr
Vehicle(s) cost at .40/mile	8.00	20 miles round trip
Cost for one delivery	33.10	
% of crop to total load	10%	for example
x number of trips	12	for example
Delivery cost for crop per season:	**39.72**	

Delivery costs can be determined for each trip, total trips per season, or the percentage cost of each product delivered. If a delivery contains equal amounts of carrots and beets, 50% of the delivery cost would be allotted to each crop.

Farmers' Market Costs:

		Calculate for ONE market
Labor: load truck(s)	12.55	1 hr (2 people @.5 hr each)
Labor: travel to market, set up	50.20	4 hrs (2 people)
Labor: market vending	100.40	8 hrs (2 people)
Labor: pack up, travel home, unpack, tally sales	37.65	3 hrs (2 people)
Vehicle(s) cost at .40/mile	8.00	20 miles round trip
Rental fees	30.00	per market
Amortized FM equipment	7.67	scales $1500, umbrellas $400, tables $200, signs $200 = $2300/15-year useful life/20 markets per season = $7.67 per market
Subtotal, cost for one market:	**246.47**	
# of markets where crop is sold	6	varies by crop
Total costs for # of markets	1478.82	
Crop sales/total FM sales	5%	varies by crop
Crop sales % x total market costs:	**73.94**	Enter in Crop Enterprise Budget under "Marketing Costs: Farmers' market expense"

The base cost for attending one market is constant irrespective of the amount of product sold (unless labor needs change). Gross sales at market must be higher than the cost; otherwise, you are losing money or personally subsidizing the market cost by not paying yourself the going labor rate. Sales need to be high enough to justify the cost of vending at market. If they are not, strive for higher sales or pursue alternative selling venues, such as CSA programs or wholesale accounts.

The total expense for equipment needed at market is amortized over the useful life of the equipment and prorated for each market. As with delivery costs above, a percentage of farmers' market expense can be assigned to different crops. The important message regarding farmers' market costs, though, is that each market costs a certain amount to attend, and that farmers' market sales must justify that expense.

Overhead Costs (annual)

Overhead costs are ones not accounted for in delivery costs, farmers' market costs, greenhouses, tractors, implement, or irrigation costs. Overhead costs are spread out over the entire farm operation and prorated to each crop or enterprise. In these worksheets, 75% of overhead expenses are apportioned to the 5 acres in cultivation, 12.5% to the bedding-plant greenhouse, and 12.5% to the in-ground tomato greenhouse. Allotment of overhead costs is somewhat subjective, but all overhead costs must be assigned. Overhead expenses allotted to the cultivated 5 acres is further broken down to overhead expense per two 350'-long beds, the equivalent of 1/10 acre.

Mortgage annual payment	600.00	farm % of total bill. Does not include house and house site portion.
Depreciation	2000.00	to account for replacement costs, excluding machinery in Worksheet 4
Property taxes	800.00	farm %
Insurance	4000.00	$3000 health, $1000 fire; not vehicle or workers' comp.
Office	1100.00	supplies, postage, subscriptions
Website	400.00	$20/month plus fees and maintenance
Travel/conferences	300.00	
Professional services	700.00	CPA, organic certification, snowplowing
Electric	600.00	farm %, w/o greenhouse electrical use
Landfill	250.00	
Telephone	550.00	farm %
Advertising	200.00	
Shop supplies, misc. repairs	500.00	tractor, implement, irrigation repairs already accounted for in Worksheet 4
Labor: management	3263.00	average 5 hrs/week, 260 hrs/year; annual labor for overseeing farm operation
Labor: office	3263.00	average 5 hrs/week, 260 hrs/year; annual labor for office duties
Labor: maintenance	653.00	average 1 hr/week, 52 hrs/year; annual labor for nonassigned maintenance work
Total overhead costs:	**19179.00**	**Allocation:** GH seedlings $2397, GH tomatoes $2397, 5A (100 beds) $14,385 = $144 per bed
Overhead per two 350' beds:	**288.00**	Per two 350' beds, for 5A (100 beds) planted to row crops. Enter on line 69 on Crop Enterprise Budget.
Overhead per greenhouse:	**2397.00**	Per 21' x 96' hoophouse: one for bedding plants, one for greenhouse tomatoes

Worksheet 2

Worksheet 2

Greenhouse Flat Costs for Calculating Worksheet 3 Bedding-Plant Cost

Costs of Soil, Plastic Containers, and Labor Filling

In order to calculate what a farm-raised seedling costs, we first need to know the cost of the plastic container, the cost of the soil in the container, and the cost of labor to fill the container. Below is a table that lists common pack sizes used in greenhouse production and the associated costs with that size. A 1020 is a 10" x 20" open plastic tray. One 1020 tray will hold eighteen 3.5" square pots. A 606 is six 6-packs sized to fit a 1020 tray. An 804 is eight 4-packs sized to fit a 1020 tray. An 806 is a eight 6-packs sized to fit a 1020 tray. 128 and 98 stand for the number of molded individual cells in a 1020-sized tray. Reuse of plastic containers will lower costs.

	A	B	C	D: C/B	E	F	G: F/G	H: A + D + G
Container size	Single-use cost/flat	# of containers per yard of soil	Price per yard of soil	Cost of soil in container	# of flats filled per hour	Labor cost per hour	Cost of labor to fill flat	Total cost of plastic, soil, and labor (w/o 1020)
3.5" square pot (18/tray)	1.62	125	105	0.84	40	12.55	0.31	2.77
606	0.39	144	105	0.73	60	12.55	0.21	1.32
804	0.39	144	105	0.73	60	12.55	0.21	1.32
806	0.39	171	105	0.61	60	12.55	0.21	1.21
1020	0.72	100	105	1.05	60	12.55	0.21	1.98
128	0.95	216	105	0.49	60	12.55	0.21	1.64
98	0.95	216	105	0.49	60	12.55	0.21	1.64
6" pot: each pot	0.28	350	105	0.30	240	12.55	0.05	0.63

Worksheet 3

Worksheet 3 Greenhouse Costs

Two types of greenhouse operations are portrayed: one for growing bedding plants and one for growing in-ground tomatoes. Both greenhouses are 21' x 96' hoop houses with two layers of plastic that are inflated. Each has a furnace, exhaust fan, intake shutters, and automatic controls. The longer-lived structure and equipment costs are totaled and divided by their useful life (20 years). Annual costs of heating fuel, electricity, and 5-year plastic covers are listed separately. Overhead expenses from Worksheet 1 (12.5% of total overhead) are added in after the annual expense subtotal. The bedding-plant greenhouse is more involved and listed first. The bedding-plant greenhouse benches hold 1000 flats (1020 size), and two flats can occupy the same bench space during the course of the bedding-plant season (one cycling of inventory). Worksheet 2 lists costs for plastic containers, soil, and the labor to fill the containers, as shown under *Production costs per flat*. Other production costs per flat are listed, with optional categories like thinning and fertilizing left blank for simplicity. The total cost per flat is a very useful number and will be used in the Crop Enterprise Budgets when crops are raised from transplants.

Bedding Plants, March 1st Start-up

Structure cost: 21' x 96', 2-layer poly-covered hoop house	
Frame cost $2400, installation $1004 (80 hrs), wood $300	3704.00
Furnace $2000, fans $800, installation $377 (30 hrs)	3177.00
Benches $500, plumbing $400, irrigation $400	1300.00
Total structure cost	8181.00
divide by # years of useful life	20
Annual structure cost	**409.05**
Other annual expenses:	
Poly cost $600, installation $100 (8 hrs), /5 years	140.00
Electricity 5 x $15/month	75.00
Fuel for heat 300 gallons @ $3/gallon	900.00
Watering labor 2 hrs x 50 times = 100 hrs	1255.00
Subtotal annual expenses	2370.00
Farm overhead allocation from Worksheet 1	2397.00
***Total annual expenses** with overhead allotment:*	**5176.05**

Greenhouse 1020 capacity: 1000 x 2	2000 one cycling of bench space		
Total annual expense/total flats =	2.59 per flat		
Greenhouse annual cost/flat:	2.59	2.59	2.59
Production costs per flat:	***804s***	***3.5" sq. pots***	***128s***
Cost of plastic flat, soil, labor filling	1.32	2.77	1.64
Cost of seed in flat	1.00	1.00	1.00
Labor to seed flat:12 flats/hr = $1.05/flat	1.05	1.05	1.05
If needed:subtotal/# of finished trays			
Labor: transplant to one flat: 10 flats/hr = $1.26			
2nd plastic flat, soil, labor filling			
Subtotal for transplanted flat			
Labor moving: 60 flats/hr = $0.21/flat each move	0.21	0.21	0.21
Labor to thin: 100 flats/hr = $0.13/flat			
Fertilizer cost:$0.02/flat			
Fertilizer labor: $0.05/flat			
Total cost per flat:	6.17	7.62	6.49

Greenhouse Tomatoes, Transplanted in Ground April 1 in Northern U.S.

The annual structure cost and other annual expenses are similar to those of the bedding-plant greenhouse shown above. Overhead costs from Worksheet 1 (12.5% of total overhead) are added in after total annual expenses. This greenhouse is used to grow tomatoes in the ground for an early and extended harvest of top-quality fruit. Tomato plants are transplanted from 3.5" pots into the greenhouse soil around April 1st. Plants are irrigated with drip lines on a battery-operated water timer. The ground is mulched to reduce weeding labor. Heating and venting are on thermostatic controls. Roll-up sidewalls promote airflow when outside temperatures permit. Tomato plants are trellised from strings hanging from the greenhouse frame. A separate crop budget is calculated for greenhouse tomatoes, shown in the Crop Enterprise Budget section. The total annual expense seen below will be used as an expense in the Crop Enterprise Budget.

Structure cost: 21' x 96' two-layer poly-covered hoop house	
Frame cost $2400, installation $1004, wood $300	3704.00
Furnace $2000, fans $800, installation $377 (30 hrs)	3177.00
Total structure cost	6881.00
Annual structure cost: divide by 20 years	344.05

Other annual expenses:	
Poly cost $600, installation $100 (8 hrs), /5 years	140.00
Electricity 6 x $15/month	90.00
Fuel for heat 200 gallons @ $3/gallon	600.00
Subtotal annual expenses	830.00
Farm overhead allocation from Worksheet 1	2397.00

Worksheet 4

Worksheet 4 Tractor, Implement, and Irrigation Costs

Tractor Costs

The hourly cost of a tractor is calculated by first dividing the purchase price of the tractor by the tractor's years of useful life. Next, annual expenses for repairs and fuel are added in, giving you the total cost to own and operate the tractor per year. Divide this total annual cost by the number of hours the tractor runs in a year, and the result is an average cost per tractor hour. I was surprised at first at how inexpensive running a tractor can be, but remember, a tractor used 50 hours per year has a much higher hourly rate than a tractor used 300 hours per year. The three tractors shown below are ones that I have owned, and the numbers are based on personal experience. Annual repairs are listed as an average: some years are expensive, some are not.

Tractor model	JD 2240	Ford 4000	Cub	
Original cost/useful life	*7000/25*	*4400/25*	*1000/25*	
Annual cost, w/o interest	280.00	176.00	40.00	
Average annual repairs	500.00	300.00	200.00	some years $0, some lots
Annual fuel cost @ $3/gallon	480.00	480.00	80.00	
Total annual cost	1260.00	956.00	320.00	
Hours used annually	200	300	60	
Tractor cost/hour	6.30	3.19	5.33	
Tractor driver hourly rate	12.55	12.55	12.55	
Tractor with driver: $/hour	18.85	15.74	17.88	

Implement Costs

Tracking various implements' costs is similar to tracking costs of tractors but without the fuel expense. Some implements have lots of moving parts (e.g., combines, manure spreaders) and cost more to operate than implements like a bedlifter, which has no moving parts. I list three of the more common and costly implements to run. Because a farm may have numerous implements, I make a note below these three implement costs for easy calculations to use as a shortcut for budget work.

	PTOTiller	Manure Spreader	Brush Hog
Original cost/useful life	*800/25*	*1100/20*	*600/20*
Annual cost, w/o interest	32.00	55.00	30.00
Implement annual repairs, average	20.00	20.00	20.00
Annual hours used	40	20	50
Implement cost/hour	1.30	3.75	1.00

A $500 simpler implement with a useful life of 25 years costs about $20/year to own. Figure $.50/hour for quick calculating.
A $1000 simpler implement with a useful life of 25 years costs about $40/year to own. Figure $1/hour for quick calculating.

Irrigation Costs

Irrigation costs take into account the annual equipment cost and any repair expense (similar to tractors and implements) and also time for setting up, running, and taking down (or moving) the system, calculated for the area that is watered each time. The example below shows an irrigation system that waters an acre in area and is used four times per season. The irrigation cost per acre is then calculated for 1/10 of an acre, or two 350'-long beds.

Cost of pipe, pump, sprinklers	4600.00	used PTO (power take-off) pump, 4" and 2" aluminum pipe for 1 acre
Useful life in years	25	
Annual equipment cost	184.00	
Average annual repairs	50.00	say $250 every 5 years
Total annual cost	234.00	
Total annual cost/uses per season	58.50	4 uses per season
Setup, takedown labor per irrigation area	75.30	1A coverage, 6 hrs total @ $12.55/hr
4 hours tractor use	25.20	at $6.30/hr, tractor only
Irrigation costs/irrigated area, each use	159.00	per acre
Irrigation costs for two 350' beds, each use	15.90	$7.53 labor, $8.37 machinery

Crop Enterprise Budget: Basil (bunches)

Crop Enterprise Budget

Crop Year: | Crop: Basil: bunches (and specify: early, mid, late) | Unit Area: Two 350' beds | Note: Twenty 350' beds = 1 acre

Bed feet or acres: 700' or 1/10A

Today's Date: | Rows per bed & plant spacing: 3 rows/bed, 12" apart on plastic

Costs in $: Remember to prorate to unit area | Field:

	$ Labor cost	$ Machinery cost	$ Product cost	NOTES: Labor at $12.55/hr. See Worksheet 1. Figures below are for two 350' beds.
Prepare Soil:				
Disk 1x	1.26	0.73		1A at a time: 1 hr total for 20 beds = 6 mins/2 beds; $1.26L, $0.63 + .10 = $0.73M w/ JD 2240; see Worksheet 4
Chisel 1x	2.51	0.74		.5A at a time:1 hr total for 10 beds = 12 mins/2 beds; $2.51L, $0.64 +.10 = $0.74M w/ Ford 4000; see Worksheet 4
Rototill 1x, 2x				.5A at a time: 2 hrs total for 10 beds = 24 mins/2 beds; $5.02L, $1.28 tractor + .52 tiller = $1.80M w/ Ford 4000
Bedform 2x	5.02	1.48		.5A at a time: 1 hr total for 10 beds = 12 mins/2 beds; $2.51L, $0.64 +.10 = $0.74M for ONE pass w/ Ford 4000
Fertilizer	1.26	0.68	10.00	500 lbs 4-3-3/A at a time: 1 hour total for 20 beds = 6 mins/2 beds; $.1.26L, $0.63 +.05 = $0.68M, $10Pr w/ JD 2240
Manure, compost	2.52	1.02	25.00	1A at a time: compost at $25/yd, 10 yds/A; 2 hrs total for 20 beds = 12 mins per 2 beds; $2.51L, $1.26 + .75 = $2.01M, $25Pr w/ JD 2240
Other				
Plastic mulch	2.09	0.70	20.00	.5A at a time: 1.5 hr/A laying = 10 mins/2 beds; $2.09L, $0.53 +.17 = $0.70M, $20Pr w/ Ford 4000
Seed/Transplant:				
Seeding in field				2 beds at a time: 30 mins/2 beds total = $6.28L
Cost of transplants			126.00	$6.49/128 = $0.06/plant 2100 plants
Transplanting labor	37.65			3 rows by hand: 3 hrs/2 beds total = $37.65L
				2 rows w/ transplanter, 6 beds at a time; 1 hr prep plants, 1.5hr x 3 people transplanting, 2 hrs machinery for 2 beds = $22.78L, $2.11 + .66 = $2.77
Cultivation:				
Reemay on/off	18.82		70.00	For 2 beds: $105/3 uses = $35Pr, .75 hr laying = $9.41L two times
Hoeing 1x, 2x, 3x	25.10			at $12.55/hr: average 1 hr/2 beds $12.55/2 beds hoe sides of plastic mulch 2x
Hand weeding 1	12.55			at $12.55/hr: average 8 hrs/2 beds $100.40/2 beds
Hand weeding 2	12.55			at $12.55/hr: average 4 hrs/2 beds $50.20/2 beds
Hand weeding 3				at $12.55/hr: average 2 hrs/2 beds $25.10/2 beds
Straw mulch				40 bales at $3, 1 hr/2 beds; $12.55L, $120.00Pr
Irrigating 1x	7.53	8.37		$7.53L, $8.37M per 2 beds, each use, w/ JD 2240
Tractor cultivating 6x	7.56	3.48		1A at a time: 1 hour/A = 6 mins/2 beds; $1.26L, $0.53 +.05 = $0.58M per pass w/ Cub mostly
Side-dressing				Spin 500 lbs 4-3-3/A, 1 hr total/20 beds = 6 mins/2 beds; $1.26L, $0.32 +.05 = $0.37M, $10Pr w/ Ford 4000
Spraying				1 hr/.5A total time = 12 mins/2 beds; $2.51L, $0.64 +.10 = $0.74M, $6Pr w/ Ford 4000
Flame weeding				10 beds/hr = 12 mins/2 beds; $2.51L, $0.64 +.10 = $0.74M, $6Pr w/ Ford 4000
Other				
Pre-harvest Subtotal:	136.42	17.20	251.00	= 404.62 Pre-harvest cost for two beds
Harvest:	Total yield for two 350' beds =			4000 bunches
	Total hours to harvest two 350' beds			66.7 hrs at 60 bunches/hr
Field to pack house	837.09			at $12.55/hr 66.7 hrs
Pack house to cooler	84.09			at $12.55/hr at 600 bunches/hr: 6.7 hrs
Bags, boxes, labels			178.69	$0.25/bag, $1.00/box, $0.07/label 24-count box = 167 at $1.07
Delivery	30.12	9.60		See Worksheet 1.
Post Harvest:				
Mow crop	2.09	0.70		6 beds at a time: 10 mins/2 beds; $2.09L, $0.53 +.17 = $0.70M w/ Ford 4000
Remove mulch	12.55			1 hour/2 beds: $12.55L
Disk	1.26	0.73		$1.26L, $0.63 +.10 = $0.73M w/ JD 2240, see disking above.
Sow cover crop: spinner	1.26	0.68	8.00	1A at a time: 1 hr/20 beds = 6 mins/2 beds; $1.26L, $0.63 + .05 = $0.68M, $8Pr w/ JD 2240
Sow cover crop: Brillion				1A at a time: 2 hrs/20 beds = 12 mins/2 beds; $2.51L, $1.26 + .20 = $1.46M, 8Pr w/ JD 2240
Other				
Post-harvest Subtotal:	1104.88	28.91	437.69	= 1571.48 Harvested cost for 2 beds
Marketing Costs:				
Labor: sales calls for season (for this crop only)	6.28			Average 10 mins/week for 3 weeks: .5 hr
Commissions				Commissions, if any, to growers' co-op, broker, or salesperson
Farmers' market expense	60.24	4.70	9.00	See Worksheet 1.
Total Crop Costs:	1171.40	33.61	446.69	= 1651.70 Total crop costs
Overhead Costs:	288.00			Apportionment for two 350' beds, see Worksheet 1.
Total Costs: Crop & Overhead Total:	1939.70			Total costs per two 350' beds

Sales:	# of units	Price per unit	Total $	
Retail:	400.00	2.50	1000.00	
Wholesale:	3600.00	1.25	4500.00	
Other:			0.00	
Total units	4000.00			
Total Sales:			5500.00	For two 350' beds

Net Profit: Total sales – total costs =	3560.30	**Net profit for two 350' beds (1/10 acre)**
Net Profit/Acre:	35603.00	Standardize to one acre
Cost/Unit:	0.48	Total cost/total units
Net Profit/Unit:	0.89	Net profit/total units

NOTES:

Crop Enterprise Budget: Beans (bush)

Crop Enterprise Budget

Crop Year:		Crop: (and specify: early, mid, late)	Beans: bush
Unit Area:	Two 350' beds	Note: Twenty 350' beds = 1 acre	
Bed feet or acres:	700' or 1/10A		
Today's Date:		Rows per bed & plant spacing:	2 rows/bed
Field:			

Costs in $: Remember to prorate to unit area

NOTES: Labor at $12.55/hr. See Worksheet 1. Figures below are for two 350' beds.

	$ Labor cost	$ Machinery cost	$ Product cost	Notes
Prepare Soil:				
Disk 1x	1.26	0.73		1A at a time: 1 hr total for 20 beds = 6 mins/2 beds; $1.26L, $0.63 + .10 = $0.73M w/ JD 2240; see Worksheet 4
Chisel 1x	2.51	0.74		.5A at a time:1 hr total for 10 beds = 12 mins/2 beds; $2.51L, $0.64 +.10 = $0.74M w/ Ford 4000; see Worksheet 4
Rototill 1x, 2x				.5A at a time: 2 hrs total for 10 beds = 24 mins/2 beds; $5.02L, $1.28 tractor + .52 tiller = $1.80M w/ Ford 4000
Bedform 2x	5.02	1.48		.5A at a time: 1 hr total for 10 beds = 12 mins/2 beds; $2.51L, $0.64 +.10 = $0.74M for ONE pass w/ Ford 4000
Fertilizer	1.26	0.68	10.00	500 lbs 4-3-3/A at a time: 1 hour total for 20 beds = 6 mins/2 beds; $.1.26L, $0.63 +.05 = $0.68M, $10Pr w/ JD 2240
Manure, compost	2.52	1.02	25.00	1A at a time: compost at $25/yd, 10 yds/A; 2 hrs total for 20 beds = 12 mins per 2 beds; $2.51L, $1.26 + .75 = $2.01M, $25Pr w/ JD 2240
Other				
Plastic mulch				.5A at a time: 1.5 hr/A laying = 10 mins/2 beds; $2.09L, $0.53 +.17 = $0.70M, $20Pr w/ Ford 4000
Seed/Transplant:				
Seeding in field	6.28		15.00	2 beds at a time: 30 mins/2 beds total = $6.28L 9 lbs seed
Cost of transplants				$6.49/128 = $0.06/plant
Transplanting labor				3 rows by hand: 3 hrs/2 beds total = $37.65L 2 rows w/ transplanter, 6 beds at a time; 1 hr prep plants, 1.5hr x 3 people transplanting, 2 hrs machinery for 2 beds = $22.78L, $2.11 + .66 = $2.77M
Cultivation:				
Reemay on/off				For 2 beds: $105/3 uses = $35Pr, .75 hr laying = $9.41L
Hoeing 1x, 2x, 3x				at $12.55/hr: average 1 hr/2 beds $12.55/2 beds
Hand weeding 1	50.20			at $12.55/hr: average 8 hrs/2 beds $100.40/2 beds easy to weed
Hand weeding 2	25.10			at $12.55/hr: average 4 hrs/2 beds $50.20/2 beds
Hand weeding 3				at $12.55/hr: average 2 hrs/2 beds $25.10/2 beds
Straw mulch				40 bales at $3, 1 hr/2 beds; $12.55L, $120.00Pr
Irrigating 1x	7.53	8.37		$7.53L, $8.37M per 2 beds, each use, w/ JD 2240
Tractor cultivating 6x	7.56	3.48		1A at a time: 1 hour/A = 6 mins/2 beds; $1.26L, $0.53 +.05 = $0.58M per pass w/ Cub mostly
Side-dressing				Spin 500 lbs 4-3-3/A, 1 hr total/20 beds = 6 mins/2 beds; $1.26L, $0.32 +.05 = $0.37M, $10Pr w/ Ford 4000
Spraying				1 hr/.5A total time = 12 mins/2 beds; $2.51L, $0.64 +.10 = $0.74M, $6Pr w/ Ford 4000
Flame weeding				10 beds/hr = 12 mins/2 beds; $2.51L, $0.64 +.10 = $0.74M, $6Pr w/ Ford 4000
Other				
Pre-harvest Subtotal:	109.24	16.50	50.00	= 175.74 Pre-harvest cost for two beds

Harvest:

Total yield for two 350' beds =	500 lbs	
Total hours to harvest two 350' beds	42 hrs	Average over 3 pickings: 12 lbs/hr

	Labor cost	Machinery cost	Product cost	Notes
Field to pack house	527.10			at $12.55/hr 42 hrs
Pack house to cooler	25.10			at $12.55/hr 2 hrs
Bags, boxes, labels			21.40	$0.25/bag, $1.00/box, $0.07/label 20 boxes at $1.07
Delivery	30.12	9.60		See Worksheet 1.
Post Harvest:				
Mow crop	2.09	0.70		6 beds at a time: 10 mins/2 beds; $2.09L, $0.53 +.17 = $0.70M w/ Ford 4000
Remove mulch				1 hour/2 beds: $12.55L
Disk	1.26	0.73		$1.26L, $0.63 +.10 = $0.73M w/ JD 2240, see disking above.
Sow cover crop: spinner	1.26	0.68	8.00	1A at a time: 1 hr/20 beds = 6 mins/2 beds; $1.26L, $0.63 + .05 = $0.68M, $8Pr w/ JD 2240
Sow cover crop: Brillion				1A at a time: 2 hrs/20 beds = 12 mins/2 beds; $2.51L, $1.26 + .20 = $1.46M, 8Pr w/ JD 2240
Other				
Post-harvest Subtotal:	696.17	28.21	79.40	= 803.78 Harvested cost for 2 beds
Marketing Costs:				
Labor: sales calls for season (for this crop only)	6.28			Average 10 mins/week for 3 weeks: .5 hr
Commissions				Commissions, if any, to growers' co-op, broker, or salesperson
Farmers' market expense	60.24	4.70	9.00	See Worksheet 1.
Total Crop Costs:	762.69	32.91	88.40	= 884.00 Total crop costs
Overhead Costs:	288.00			Apportionment for two 350' beds, see Worksheet 1.
Total Costs: Crop & Overhead Total:	1172.00			Total costs per two 350' beds

Sales:	# of units	Price per unit	Total $	
Retail:	250.00	2.50	625.00	
Wholesale:	250.00	1.10	275.00	
Other:			0.00	
Total units	500.00			
Total Sales:			900.00	For two 350' beds

Net Profit: Total sales – total costs =	-272.00	**Net profit for two 350' beds (1/10 acre)**
Net Profit/Acre:	-2720.00	Standardize to one acre
Cost/Unit:	2.34	Total cost/total units
Net Profit/Unit:	-0.54	Net profit/total units

NOTES:

Crop Enterprise Budget: Beets (roots)

Crop Enterprise Budget

Crop Year: | Crop: **Beets: roots** (and specify: early, mid, late) | **Unit Area:** **Two 350' beds** | **Note: Twenty 350' beds = 1 acre**

Bed feet or acres: **700' or 1/10A**

Today's Date: | Rows per bed & plant spacing: 3 rows/bed, 15 seeds/foot

Costs in $: | Remember to prorate to unit area | Field:

NOTES: Labor at $12.55/hr. See Worksheet 1. Figures below are for two 350' beds.

	Labor cost $	Machinery cost $	Product cost $	Notes
Prepare Soil:				
Disk 1x	1.26	0.73		1A at a time: 1 hr total for 20 beds = 6 mins/2 beds; $1.26L, $0.63 + .10 = $0.73M w/ JD 2240; see Worksheet 4
Chisel 1x	2.51	0.74		.5A at a time:1 hr total for 10 beds = 12 mins/2 beds; $2.51L, $0.64 +.10 = $0.74M w/ Ford 4000; see Worksheet 4
Rototill 1x, 2x				.5A at a time: 2 hrs total for 10 beds = 24 mins/2 beds; $5.02L, $1.28 tractor + .52 tiller = $1.80M w/ Ford 4000
Bedform 2x	5.02	1.48		.5A at a time: 1 hr total for 10 beds = 12 mins/2 beds; $2.51L, $0.64 +.10 = $0.74M for ONE pass w/ Ford 4000
Fertilizer	1.26	0.68	10.00	500 lbs 4-3-3/A at a time: 1 hour total for 20 beds = 6 mins/2 beds; $.1.26L, $0.63 +.05 = $0.68M, $10Pr w/ JD 2240
Manure, compost	2.52	1.02	25.00	1A at a time: compost at $25/yd, 10 yds/A; 2 hrs total for 20 beds = 12 mins per 2 beds; $2.51L, $1.26 + .75 = $2.01M, $25Pr w/ JD 2240
Other				
Plastic mulch				.5A at a time: 1.5 hr/A laying = 10 mins/2 beds; $2.09L, $0.53 +.17 = $0.70M, $20Pr w/ Ford 4000
Seed/Transplant:				
Seeding in field	6.28		53.00	2 beds at a time: 30 mins/2 beds total = $6.28L 32,000 seeds
Cost of transplants				$6.49/128 = $0.06/plant
Transplanting labor				3 rows by hand: 3 hrs/2 beds total = $37.65L
				2 rows w/ transplanter, 6 beds at a time; 1 hr prep plants, 1.5hr x 3 people transplanting, 2 hrs machinery for 2 beds = $22.78L, $2.11 + .66 = $2.77M
Cultivation:				
Reemay on/off				For 2 beds: $105/3 uses = $35Pr, .75 hr laying = $9.41L
Hoeing 1x, 2x, 3x				at $12.55/hr: average 1 hr/2 beds $12.55/2 beds
Hand weeding 1	100.40			at $12.55/hr: average 8 hrs/2 beds $100.40/2 beds
Hand weeding 2	50.20			at $12.55/hr: average 4 hrs/2 beds $50.20/2 beds
Hand weeding 3				at $12.55/hr: average 2 hrs/2 beds $25.10/2 beds
Straw mulch				40 bales at $3, 1 hr/2 beds; $12.55L, $120.00Pr
Irrigating 1x	7.53	8.37		$7.53L, $8.37M per 2 beds, each use, w/ JD 2240
Tractor cultivating 6x	7.56	3.48		1A at a time: 1 hour/A = 6 mins/2 beds; $1.26L, $0.53 +.05 = $0.58M per pass w/ Cub mostly
Side-dressing				Spin 500 lbs 4-3-3/A, 1 hr total/20 beds = 6 mins/2 beds; $1.26L, $0.32 +.05 = $0.37M, $10Pr w/ Ford 4000
Spraying				1 hr/.5A total time = 12 mins/2 beds; $2.51L, $0.64 +.10 = $0.74M, $6Pr w/ Ford 4000
Flame weeding				10 beds/hr = 12 mins/2 beds; $2.51L, $0.64 +.10 = $0.74M, $6Pr w/ Ford 4000
Other				
Pre-harvest Subtotal:	184.54	16.50	88.00	= 289.04 Pre-harvest cost for two beds

Harvest:

Total yield for two 350' beds = 60 bags

Total hours to harvest two 350' beds 10 hrs at 6 bags/hr

	Labor cost $	Machinery cost $	Product cost $	Notes
Field to pack house	125.50			at $12.55/hr 10 hours
Pack house to cooler	75.30			at $12.55/hr 10 bags/hr washing = 6 hrs
Bags, boxes, labels			15.00	$0.25/bag, $1.00/box, $0.07/label
Delivery	30.12	9.60		See Worksheet 1.
Post Harvest:				
Mow crop				6 beds at a time: 10 mins/2 beds; $2.09L, $0.53 +.17 = $0.70M w/ Ford 4000
Remove mulch				1 hour/2 beds: $12.55L
Disk	1.26	0.73		$1.26L, $0.63 +.10 = $0.73M w/ JD 2240, see disking above.
Sow cover crop: spinner	1.26	0.68	8.00	1A at a time: 1 hr/20 beds = 6 mins/2 beds; $1.26L, $0.63 + .05 = $0.68M, $8Pr w/ JD 2240
Sow cover crop: Brillion				1A at a time: 2 hrs/20 beds = 12 mins/2 beds; $2.51L, $1.26 + .20 = $1.46M, 8Pr w/ JD 2240
Other				
Post-harvest Subtotal:	417.98	27.51	111.00	= 556.49 Harvested cost for 2 beds
Marketing Costs:				
Labor: sales calls for season (for this crop only)	6.28			Average 10 mins/week for 3 weeks: .5 hr
Commissions				Commissions, if any, to growers' co-op, broker, or salesperson
Farmers' market expense	60.24	4.70	9.00	See Worksheet 1.
Total Crop Costs:	484.50	32.21	120.00	= 636.71 Total crop costs
Overhead Costs:	288.00			Apportionment for two 350' beds, see Worksheet 1.
Total Costs: Crop & Overhead Total:	924.71			Total costs per two 350' beds

Sales:	# of units	Price per unit	Total $
Retail:	10.00	50.00	500.00
Wholesale:	50.00	25.00	1250.00
Other:			0.00
Total units	60.00		
Total Sales:			1750.00 For two 350' beds

Net Profit: Total sales – total costs =	825.29	**Net profit for two 350' beds (1/10 acre)**
Net Profit/Acre:	8252.90	Standardize to one acre
Cost/Unit:	15.41	Total cost/total units
Net Profit/Unit:	13.75	Net profit/total units

NOTES:

Crop Enterprise Budget: Broccoli

Crop Enterprise Budget

Crop Year: | Crop: **Broccoli** (and specify: early, mid, late) | Unit Area: **Two 350' beds** | Bed feet or acres: **700' or 1/10A** | **Note: Twenty 350' beds = 1 acre**

Today's Date: | Rows per bed & plant spacing: 2 rows/bed, 12" transplant spacing | Field:

Costs in $: Remember to prorate to unit area

	$ Labor cost	$ Machinery cost	$ Product cost	NOTES: Labor at $12.55/hr. See Worksheet 1. Figures below are for two 350' beds.
Prepare Soil:				
Disk 1x	1.26	0.73		1A at a time: 1 hr total for 20 beds = 6 mins/2 beds; $1.26L, $0.63 + .10 = $0.73M w/ JD 2240; see Worksheet 4
Chisel 1x	2.51	0.74		.5A at a time:1 hr total for 10 beds = 12 mins/2 beds; $2.51L, $0.64 +.10 = $0.74M w/ Ford 4000; see Worksheet 4
Rototill 1x, 2x				.5A at a time: 2 hrs total for 10 beds = 24 mins/2 beds; $5.02L, $1.28 tractor + .52 tiller = $1.80M w/ Ford 4000
Bedform 2x	5.02	1.48		.5A at a time: 1 hr total for 10 beds = 12 mins/2 beds; $2.51L, $0.64 +.10 = $0.74M for ONE pass w/ Ford 4000
Fertilizer	1.26	0.68	10.00	500 lbs 4-3-3/A at a time: 1 hour total for 20 beds = 6 mins/2 beds; $.1.26L, $0.63 +.05 = $0.68M, $10Pr w/ JD 2240
Manure, compost	2.52	1.02	25.00	1A at a time: compost at $25/yd, 10 yds/A; 2 hrs total for 20 beds = 12 mins per 2 beds; $2.51L, $1.26 + .75 = $2.01M, $25Pr w/ JD 2240
Other				
Plastic mulch				.5A at a time: 1.5 hr/A laying = 10 mins/2 beds; $2.09L, $0.53 +.17 = $0.70M, $20Pr w/ Ford 4000
Seed/Transplant:				
Seeding in field				2 beds at a time: 30 mins/2 beds total = $6.28L
Cost of transplants			84.00	$6.49/128 = $0.06/plant — 1400 plants at $0.06
Transplanting labor	25.23			3 rows by hand: 3 hrs/2 beds total = $37.65L — only 2 rows/bed
				2 rows w/ transplanter, 6 beds at a time; 1 hr prep plants, 1.5hr x 3 people transplanting, 2 hrs machinery for 2 beds = $22.78L, $2.11 + .66 = $2.77M
Cultivation:				
Reemay on/off				For 2 beds: $105/3 uses = $35Pr, .75 hr laying = $9.41L
Hoeing 1x, 2x, 3x	12.55			at $12.55/hr: average 1 hr/2 beds — $12.55/2 beds
Hand weeding 1	25.10			at $12.55/hr: average 8 hrs/2 beds — $100.40/2 beds
Hand weeding 2				at $12.55/hr: average 4 hrs/2 beds — $50.20/2 beds
Hand weeding 3				at $12.55/hr: average 2 hrs/2 beds — $25.10/2 beds
Straw mulch				40 bales at $3, 1 hr/2 beds; $12.55L, $120.00Pr
Irrigating 1x	7.53	8.37		$7.53L, $8.37M per 2 beds, each use, w/ JD 2240
Tractor cultivating 6x	7.56	3.48		1A at a time: 1 hour/A = 6 mins/2 beds; $1.26L, $0.53 +.05 = $0.58M per pass w/ Cub mostly
Side-dressing				Spin 500 lbs 4-3-3/A, 1 hr total/20 beds = 6 mins/2 beds; $1.26L, $0.32 +.05 = $0.37M, $10Pr w/ Ford 4000
Spraying	2.51	0.74	6.00	1 hr/.5A total time = 12 mins/2 beds; $2.51L, $0.64 +.10 = $0.74M, $6Pr w/ Ford 4000
Flame weeding				10 beds/hr = 12 mins/2 beds; $2.51L, $0.64 +.10 = $0.74M, $6Pr w/ Ford 4000
Other				
Pre-harvest Subtotal:	93.05	17.24	125.00	= 235.29 Pre-harvest cost for two beds

Harvest:

Total yield for two 350' beds = 36 cases — season average: 500 bunches, 14-count case

Total hours to harvest two 350' beds 6 hrs — 6 cases/hour

	Labor cost	Machinery cost	Product cost	Notes
Field to pack house	75.30			at $12.55/hr — 6 hrs
Pack house to cooler	37.65			at $12.55/hr — 12 cases/hour packing
Bags, boxes, labels			19.44	$0.25/bag, $1.00/box, $0.07/label — $1.07 per box/ 2 uses
Delivery	30.12	9.60		See Worksheet 1.
Post Harvest:				
Mow crop	2.09	0.70		6 beds at a time: 10 mins/2 beds; $2.09L, $0.53 +.17 = $0.70M w/ Ford 4000
Remove mulch				1 hour/2 beds: $12.55L
Disk	1.26	0.73		$1.26L, $0.63 +.10 = $0.73M w/ JD 2240, see disking above.
Sow cover crop: spinner	1.26	0.68	8.00	1A at a time: 1 hr/20 beds = 6 mins/2 beds; $1.26L, $0.63 + .05 = $0.68M, $8Pr w/ JD 2240
Sow cover crop: Brillion				1A at a time: 2 hrs/20 beds = 12 mins/2 beds; $2.51L, $1.26 + .20 = $1.46M, 8Pr w/ JD 2240
Other				
Post-harvest Subtotal:	240.73	28.95	152.44	= 422.12 Harvested cost for 2 beds
Marketing Costs:				
Labor: sales calls for season (for this crop only)	6.28			Average 10 mins/week for 3 weeks: .5 hr
Commissions				Commissions, if any, to growers' co-op, broker, or salesperson
Farmers' market expense	60.24	4.70	9.00	See Worksheet 1.
Total Crop Costs:	307.25	33.65	161.44	= 502.34 Total crop costs
Overhead Costs:	288.00			Apportionment for two 350' beds, see Worksheet 1.
Total Costs: Crop & Overhead Total:	790.34			Total costs per two 350' beds

Sales:	# of units	Price per unit	Total $
Retail:	12.00	31.50	378.00
Wholesale:	24.00	22.00	528.00
Other:			0.00
Total units	36.00		
Total Sales:			906.00 For two 350' beds

Net Profit: Total sales – total costs =	115.66	**Net profit for two 350' beds (1/10 acre)**
Net Profit/Acre:	1156.60	Standardize to one acre
Cost/Unit:	21.95	Total cost/total units
Net Profit/Unit:	3.21	Net profit/total units

NOTES:

Crop Enterprise Budget: Cabbage

Crop Enterprise Budget

Crop Year: | Crop: **Cabbage** (and specify: early, mid, late) | **Unit Area:** **Two 350' beds** | **Note: Twenty 350' beds = 1 acre**

Bed feet or acres: **700' or 1/10A**

Today's Date: | Rows per bed & plant spacing: 2 rows/bed, Tplanted 16" apart, no mulch

Costs in $: Remember to prorate to unit area | Field:

NOTES: Labor at $12.55/hr. See Worksheet 1 — Figures below are for two 350' beds

	Labor cost $	Machinery cost $	Product cost $	Notes	
Prepare Soil:					
Disk 1x	1.26	0.73		1A at a time: 1 hr total for 20 beds = 6 mins/2 beds; $1.26L, $0.63 + .10 = $0.73M w/ JD 2240; see Worksheet 4	
Chisel 1x	2.51	0.74		.5A at a time:1 hr total for 10 beds = 12 mins/2 beds; $2.51L, $0.64 +.10 = $0.74M w/ Ford 4000; see Worksheet 4	
Rototill 1x, 2x				.5A at a time: 2 hrs total for 10 beds = 24 mins/2 beds; $5.02L, $1.28 tractor + .52 tiller = $1.80M w/ Ford 4000	
Bedform 2x	5.02	1.48		.5A at a time: 1 hr total for 10 beds = 12 mins/2 beds; $2.51L, $0.64 +.10 = $0.74M for ONE pass w/ Ford 4000	
Fertilizer	1.26	0.68	10.00	500 lbs 4-3-3/A at a time: 1 hour total for 20 beds = 6 mins/2 beds; $.1.26L, $0.63 +.05 = $0.68M, $10Pr w/ JD 2240	
Manure, compost	2.52	1.02	25.00	1A at a time: compost at $25/yd, 10 yds/A; 2 hrs total for 20 beds = 12 mins per 2 beds; $2.51L, $1.26 + .75 = $2.01M, $25Pr w/ JD 2240	
Other					
Plastic mulch				.5A at a time: 1.5 hr/A laying = 10 mins/2 beds; $2.09L, $0.53 +.17 = $0.70M, $20Pr w/ Ford 4000	
Seed/Transplant:					
Seeding in field				2 beds at a time: 30 mins/2 beds total = $6.28L	
Cost of transplants			63.00	$6.49/128 = $0.06/plant	1050 plants
Transplanting labor	25.23			3 rows by hand: 3 hrs/2 beds total = $37.65L	2/3 of 3-row time
				2 rows w/ transplanter, 6 beds at a time; 1 hr prep plants, 1.5hr x 3 people transplanting, 2 hrs machinery for 2 beds = $22.78L, $2.11 + .66 = $2.77M	
Cultivation:					
Reemay on/off				For 2 beds: $105/3 uses = $35Pr, .75 hr laying = $9.41L	
Hoeing 1x, 2x, 3x	25.10			at $12.55/hr: average 1 hr/2 beds	$12.55/2 beds
Hand weeding 1	50.20			at $12.55/hr: average 8 hrs/2 beds	$100.40/2beds
Hand weeding 2	25.10			at $12.55/hr: average 4 hrs/2 beds	$50.20/2beds
Hand weeding 3				at $12.55/hr: average 2 hrs/2 beds	$25.10/2beds
Straw mulch				40 bales at $3, 1 hr/2 beds; $12.55L, $120.00Pr	
Irrigating 1x	7.53	8.37		$7.53L, $8.37M per 2 beds, each use, w/ JD 2240	
Tractor cultivating 6x	7.56	3.48		1A at a time: 1 hour/A = 6 mins/2 beds; $1.26L, $0.53 +.05 = $0.58M per pass w/ Cub mostly	
Side-dressing				Spin 500 lbs 4-3-3/A, 1 hr total/20 beds = 6 mins/2 beds; $1.26L, $0.32 +.05 = $0.37M, $10Pr w/ Ford 4000	
Spraying	2.51	0.74	6.00	1 hr/.5A total time = 12 mins/2 beds; $2.51L, $0.64 +.10 = $0.74M, $6Pr w/ Ford 4000	
Flame weeding				10 beds/hr = 12 mins/2 beds; $2.51L, $0.64 +.10 = $0.74M, $6Pr w/ Ford 4000	
Other					
Pre-harvest Subtotal:	155.80	17.24	104.00	= 277.04 Pre-harvest cost for two beds	

Harvest:

Total yield for two 350' beds = 50 cases — 900 heads: 50 18-count or 50-lb cases

Total hours to harvest two 350' beds 6.25 hrs — at 8 cases/hr

	Labor cost $	Machinery cost $	Product cost $	Notes	
Field to pack house	78.44			at $12.55/hr	6.25 hrs
Pack house to cooler	62.75			at $12.55/hr	at 10 50-lb cases/hr
Bags, boxes, labels			78.50	$0.25/bag, $1.00/box, $0.07/label	at $1.57/cabbage box
Delivery	30.12	9.60		See Worksheet 1.	
Post Harvest:					
Mow crop	2.09	0.70		6 beds at a time: 10 mins/2 beds; $2.09L, $0.53 +.17 = $0.70M w/ Ford 4000	
Remove mulch				1 hour/2 beds: $12.55L	
Disk	1.26	0.73		$1.26L, $0.63 +.10 = $0.73M w/ JD 2240, see disking above.	
Sow cover crop: spinner	1.26	0.68	8.00	1A at a time: 1 hr/20 beds = 6 mins/2 beds; $1.26L, $0.63 + .05 = $0.68M, $8Pr w/ JD 2240	
Sow cover crop: Brillion				1A at a time: 2 hrs/20 beds = 12 mins/2 beds; $2.51L, $1.26 + .20 = $1.46M, 8Pr w/ JD 2240	
Other					
Post-harvest Subtotal:	331.72	28.95	190.50	= 551.17 Harvested cost for 2 beds	
Marketing Costs:					
Labor: sales calls for season (for this crop only)	6.28			Average 10 mins/week for 3 weeks: .5 hr	
Commissions				Commissions, if any, to growers' co-op, broker, or salesperson	
Farmers' market expense	60.24	4.70	9.00	See Worksheet 1.	
Total Crop Costs:	398.24	33.65	199.50	= 631.39 Total crop costs	
Overhead Costs:	288.00			Apportionment for two 350' beds, see Worksheet 1.	
Total Costs:					
Crop & Overhead Total:	919.39			Total costs per two 350' beds	

Sales:	# of units	Price per unit	Total $	
Retail:	10.00	50.00	500.00	
Wholesale:	40.00	25.00	1000.00	
Other:			0.00	
Total units	50.00			
Total Sales:			1500.00	For two 350' beds

Net Profit: Total sales – total costs =	580.61	**Net profit for two 350' beds (1/10 acre)**
Net Profit/Acre:	5806.10	Standardize to one acre
Cost/Unit:	18.39	Total cost/total units
Net Profit/Unit:	11.61	Net profit/total units

NOTES:

Crop Enterprise Budget: Carrots (roots)

Crop Enterprise Budget

Crop Year: | Crop: **Carrots: roots** (and specify: early, mid, late) | **Unit Area:** **Two 350' beds** | **Note: Twenty 350' beds = 1 acre**

Bed feet or acres: **700' or 1/10A**

Today's Date: | Rows per bed & plant spacing: 3 rows/bed, 25 seeds/foot

Costs in $: Remember to prorate to unit area | **Field:**

NOTES: Labor at $12.55/hr. See Worksheet 1. Figures below are for two 350' beds

	$ Labor cost	$ Machinery cost	$ Product cost	Notes
Prepare Soil:				
Disk 1x	1.26	0.73		1A at a time: 1 hr total for 20 beds = 6 mins/2 beds; $1.26L, $0.63 + .10 = $0.73M w/ JD 2240; see Worksheet 4
Chisel 1x	2.51	0.74		.5A at a time:1 hr total for 10 beds = 12 mins/2 beds; $2.51L, $0.64 +.10 = $0.74M w/ Ford 4000; see Worksheet 4
Rototill 1x, 2x				.5A at a time: 2 hrs total for 10 beds = 24 mins/2 beds; $5.02L, $1.28 tractor + .52 tiller = $1.80M w/ Ford 4000
Bedform 2x	5.02	1.48		.5A at a time: 1 hr total for 10 beds = 12 mins/2 beds; $2.51L, $0.64 +.10 = $0.74M for ONE pass w/ Ford 4000
Fertilizer	1.26	0.68	10.00	500 lbs 4-3-3/A at a time: 1 hour total for 20 beds = 6 mins/2 beds; $.1.26L, $0.63 +.05 = $0.68M, $10Pr w/ JD 2240
Manure, compost	2.52	1.02	25.00	1A at a time: compost at $25/yd, 10 yds/A; 2 hrs total for 20 beds = 12 mins per 2 beds; $2.51L, $1.26 + .75 = $2.01M, $25Pr w/ JD 2240
Other				
Plastic mulch				.5A at a time: 1.5 hr/A laying = 10 mins/2 beds; $2.09L, $0.53 +.17 = $0.70M, $20Pr w/ Ford 4000
Seed/Transplant:				
Seeding in field	6.28		39.00	2 beds at a time: 30 mins/2 beds total = $6.28L — 52,000 seeds
Cost of transplants				$6.49/128 = $0.06/plant
Transplanting labor				3 rows by hand: 3 hrs/2 beds total = $37.65L 2 rows w/ transplanter, 6 beds at a time; 1 hr prep plants, 1.5hr x 3 people transplanting, 2 hrs machinery for 2 beds = $22.78L, $2.11 + .66 = $2.77M
Cultivation:				
Reemay on/off				For 2 beds: $105/3 uses = $35Pr, .75 hr laying = $9.41L
Hoeing 1x, 2x, 3x				at $12.55/hr: average 1 hr/2 beds — $12.55/2 beds
Hand weeding 1	50.20			at $12.55/hr: average 8 hrs/2 beds — $100.40/2 beds — reduced by flaming
Hand weeding 2	50.20			at $12.55/hr: average 4 hrs/2 beds — $50.20/2 beds
Hand weeding 3	25.10			at $12.55/hr: average 2 hrs/2 beds — $25.10/2 beds
Straw mulch				40 bales at $3, 1 hr/2 beds; $12.55L, $120.00Pr
Irrigating 1x	7.53	8.37		$7.53L, $8.37M per 2 beds, each use, w/ JD 2240
Tractor cultivating 6x	7.56	3.48		1A at a time: 1 hour/A = 6 mins/2 beds; $1.26L, $0.53 +.05 = $0.58M per pass w/ Cub mostly
Side-dressing				Spin 500 lbs 4-3-3/A, 1 hr total/20 beds = 6 mins/2 beds; $1.26L, $0.32 +.05 = $0.37M, $10Pr w/ Ford 4000
Spraying				1 hr/.5A total time = 12 mins/2 beds; $2.51L, $0.64 +.10 = $0.74M, $6Pr w/ Ford 4000
Flame weeding	2.51	0.74	6.00	10 beds/hr = 12 mins/2 beds; $2.51L, $0.64 +.10 = $0.74M, $6Pr w/ Ford 4000
Other				
Pre-harvest Subtotal:	161.95	17.24	80.00	= 259.19 Pre-harvest cost for two beds
Harvest:	Total yield for two 350' beds = 80 bags			
	Total hours to harvest two 350' beds 20 hrs			4 bags/hour
Field to pack house	251.00	1.85		at $12.55/hr — 20 hrs
Pack house to cooler	143.45			at $12.55/hr — 7 bags/hr
Bags, boxes, labels			20.00	$0.25/bag, $1.00/box, $0.07/label
Delivery	30.12	9.60		See Worksheet 1.
Post Harvest:				
Mow crop				6 beds at a time: 10 mins/2 beds; $2.09L, $0.53 +.17 = $0.70M w/ Ford 4000
Remove mulch				1 hour/2 beds: $12.55L
Disk	1.26	0.73		$1.26L, $0.63 +.10 = $0.73M w/ JD 2240, see disking above.
Sow cover crop: spinner	1.26	0.68	8.00	1A at a time: 1 hr/20 beds = 6 mins/2 beds; $1.26L, $0.63 + .05 = $0.68M, $8Pr w/ JD 2240
Sow cover crop: Brillion				1A at a time: 2 hrs/20 beds = 12 mins/2 beds; $2.51L, $1.26 + .20 = $1.46M, 8Pr w/ JD 2240
Other				
Post-harvest Subtotal:	589.04	30.10	108.00	= 727.14 Harvested cost for 2 beds
Marketing Costs:				
Labor: sales calls for season (for this crop only)	6.28			Average 10 mins/week for 3 weeks: .5 hr
Commissions				Commissions, if any, to growers' co-op, broker, or salesperson
Farmers' market expense	60.24	4.70	9.00	See Worksheet 1.
Total Crop Costs:	655.56	34.80	117.00	= 807.36 Total crop costs
Overhead Costs:	288.00			Apportionment for two 350' beds, see Worksheet 1.
Total Costs: Crop & Overhead Total:	1095.36			Total costs per two 350' beds

Sales:	# of units	Price per unit	Total $
Retail:	20.00	50.00	1000.00
Wholesale:	60.00	25.00	1500.00
Other:			0.00
Total units	80.00		
Total Sales:			2500.00 For two 350' beds

Net Profit: Total sales – total costs =	1404.64	**Net profit for two 350' beds (1/10 acre)**
Net Profit/Acre:	14046.40	Standardize to one acre
Cost/Unit:	13.69	Total cost/total units
Net Profit/Unit:	17.56	Net profit/total units

NOTES:
Flame weeding reduces initial weeding

Crop Enterprise Budget: Celeriac

Crop Enterprise Budget

Crop Year: | Crop: Celeriac (and specify: early, mid, late) | Unit Area: Two 350' beds | Note: Twenty 350' beds = 1 acre

Bed feet or acres: 700' or 1/10A

Today's Date: | Rows per bed & plant spacing: 2 rows/bed, Tplanted 8" in row, no mulch

Costs in $: Remember to prorate to unit area | Field:

	$ Labor cost	$ Machinery cost	$ Product cost	NOTES: Labor at $12.55/hr. See Worksheet 1. Figures below are for two 350' beds
Prepare Soil:				
Disk 1x	1.26	0.73		1A at a time: 1 hr total for 20 beds = 6 mins/2 beds; $1.26L, $0.63 + .10 = $0.73M w/ JD 2240; see Worksheet 4
Chisel 1x	2.51	0.74		.5A at a time:1 hr total for 10 beds = 12 mins/2 beds; $2.51L, $0.64 +.10 = $0.74M w/ Ford 4000; see Worksheet 4
Rototill 1x, 2x				.5A at a time: 2 hrs total for 10 beds = 24 mins/2 beds; $5.02L, $1.28 tractor + .52 tiller = $1.80M w/ Ford 4000
Bedform 2x	5.02	1.48		.5A at a time: 1 hr total for 10 beds = 12 mins/2 beds; $2.51L, $0.64 +.10 = $0.74M for ONE pass w/ Ford 4000
Fertilizer	1.26	0.68	10.00	500 lbs 4-3-3/A at a time: 1 hour total for 20 beds = 6 mins/2 beds; $.1.26L, $0.63 +.05 = $0.68M, $10Pr w/ JD 2240
Manure, compost	2.52	1.02	25.00	1A at a time: compost at $25/yd, 10 yds/A; 2 hrs total for 20 beds = 12 mins per 2 beds; $2.51L, $1.26 + .75 = $2.01M, $25Pr w/ JD 2240
Other				
Plastic mulch				.5A at a time: 1.5 hr/A laying = 10 mins/2 beds; $2.09L, $0.53 +.17 = $0.70M, $20Pr w/ Ford 4000
Seed/Transplant:				
Seeding in field				2 beds at a time: 30 mins/2 beds total = $6.28L
Cost of transplants			126.00	$6.49/128 = $0.06/plant 2100 plants
Transplanting labor	37.65			3 rows by hand: 3 hrs/2 beds total = $37.65L only 2 rows but closer spacing
				2 rows w/ transplanter, 6 beds at a time; 1 hr prep plants, 1.5hr x 3 people transplanting, 2 hrs machinery for 2 beds = $22.78L, $2.11 + .66 = $2.77M
Cultivation:				
Reemay on/off				For 2 beds: $105/3 uses = $35Pr, .75 hr laying = $9.41L
Hoeing 1x, 2x, 3x				at $12.55/hr: average 1 hr/2 beds $12.55/2 beds
Hand weeding 1	100.40			at $12.55/hr: average 8 hrs/2 beds $100.40/2 beds
Hand weeding 2	50.20			at $12.55/hr: average 4 hrs/2 beds $50.20/2 beds
Hand weeding 3	25.10			at $12.55/hr: average 2 hrs/2 beds $25.10/2 beds
Straw mulch				40 bales at $3, 1 hr/2 beds; $12.55L, $120.00Pr
Irrigating 1x	7.53	8.37		$7.53L, $8.37M per 2 beds, each use, w/ JD 2240
Tractor cultivating 6x	7.56	3.48		1A at a time: 1 hour/A = 6 mins/2 beds; $1.26L, $0.53 +.05 = $0.58M per pass w/ Cub mostly
Side-dressing	1.26	0.37	10.00	Spin 500 lbs 4-3-3/A, 1 hr total/20 beds = 6 mins/2 beds; $1.26L, $0.32 +.05 = $0.37M, $10Pr w/ Ford 4000
Spraying	2.51	0.74	6.00	1 hr/.5A total time = 12 mins/2 beds; $2.51L, $0.64 +.10 = $0.74M, $6Pr w/ Ford 4000
Flame weeding				10 beds/hr = 12 mins/2 beds; $2.51L, $0.64 +.10 = $0.74M, $6Pr w/ Ford 4000
Other	2.09	0.70		
Pre-harvest Subtotal:	246.87	18.31	177.00	= 442.18 Pre-harvest cost for two beds

Harvest:

Total yield for two 350' beds = 120 15-lb boxes

Total hours to harvest two 350' beds: 10 hrs at 12 15-lb boxes/hr

	Labor cost	Machinery cost	Product cost	Notes
Field to pack house	125.50			at $12.55/hr 10 hrs
Pack house to cooler	150.60			at $12.55/hr at 10 15-lb boxes/hr
Bags, boxes, labels			96.00	$0.25/bag, $1.00/box, $0.07/label at .80 1/2 bushel box
Delivery	30.12	9.60		See Worksheet 1.
Post Harvest:				
Mow crop				6 beds at a time: 10 mins/2 beds; $2.09L, $0.53 +.17 = $0.70M w/ Ford 4000
Remove mulch				1 hour/2 beds: $12.55L
Disk	1.26	0.73		$1.26L, $0.63 +.10 = $0.73M w/ JD 2240, see disking above.
Sow cover crop: spinner	1.26	0.68	8.00	1A at a time: 1 hr/20 beds = 6 mins/2 beds; $1.26L, $0.63 + .05 = $0.68M, $8Pr w/ JD 2240
Sow cover crop: Brillion				1A at a time: 2 hrs/20 beds = 12 mins/2 beds; $2.51L, $1.26 + .20 = $1.46M, 8Pr w/ JD 2240
Other				
Post-harvest Subtotal:	555.61	29.32	281.00	= 865.93 Harvested cost for 2 beds
Marketing Costs:				
Labor: sales calls for season (for this crop only)	6.28			Average 10 mins/week for 3 weeks: .5 hr
Commissions				Commissions, if any, to growers' co-op, broker, or salesperson
Farmers' market expense	60.24	4.70	9.00	See Worksheet 1.
Total Crop Costs:	622.13	34.02	290.00	= 946.15 Total crop costs
Overhead Costs:	288.00			Apportionment for two 350' beds, see Worksheet 1.
Total Costs: Crop & Overhead Total:	1234.15			Total costs per two 350' beds

Sales:	# of units	Price per unit	Total $
Retail:	20.00	30.00	600.00
Wholesale:	100.00	20.00	2000.00
Other:			0.00
Total units	120.00		
Total Sales:			2600.00 For two 350' beds

Net Profit: Total sales – total costs =	1365.85	**Net profit for two 350' beds (1/10 acre)**
Net Profit/Acre:	13658.50	Standardize to one acre
Cost/Unit:	10.28	Total cost/total units
Net Profit/Unit:	11.38	Net profit/total units

NOTES:

Crop Enterprise Budget: Cilantro (bunches)

Crop Enterprise Budget

Crop Year: | Crop: Cilantro: bunches (and specify: early, mid, late) | **Unit Area:** Two 350' beds | **Note: Twenty 350' beds = 1 acre**

Today's Date: | Bed feet or acres: 700' or 1/10A

Rows per bed & plant spacing: 3 rows/bed seeded thickly

Costs in $: Remember to prorate to unit area | Field:

NOTES: Labor at $12.55/hr. See Worksheet 1. Figures below are for two 350' beds.

	Labor cost $	Machinery cost $	Product cost $	Notes
Prepare Soil:				
Disk 1x	1.26	0.73		1A at a time: 1 hr total for 20 beds = 6 mins/2 beds; $1.26L, $0.63 + .10 = $0.73M w/ JD 2240; see Worksheet 4
Chisel 1x	2.51	0.74		.5A at a time:1 hr total for 10 beds = 12 mins/2 beds; $2.51L, $0.64 +.10 = $0.74M w/ Ford 4000; see Worksheet 4
Rototill 1x, 2x				.5A at a time: 2 hrs total for 10 beds = 24 mins/2 beds; $5.02L, $1.28 tractor + .52 tiller = $1.80M w/ Ford 4000
Bedform 2x	5.02	1.48		.5A at a time: 1 hr total for 10 beds = 12 mins/2 beds; $2.51L, $0.64 +.10 = $0.74M for ONE pass w/ Ford 4000
Fertilizer	1.26	0.68	10.00	500 lbs 4-3-3/A at a time: 1 hour total for 20 beds = 6 mins/2 beds; $.1.26L, $0.63 +.05 = $0.68M, $10Pr w/ JD 2240
Manure, compost	2.52	1.02	25.00	1A at a time: compost at $25/yd, 10 yds/A; 2 hrs total for 20 beds = 12 mins per 2 beds; $2.51L, $1.26 + .75 = $2.01M, $25Pr w/ JD 2240
Other				
Plastic mulch				.5A at a time: 1.5 hr/A laying = 10 mins/2 beds; $2.09L, $0.53 +.17 = $0.70M, $20Pr w/ Ford 4000
Seed/Transplant:				
Seeding in field	6.28		40.00	2 beds at a time: 30 mins/2 beds total = $6.28L — 2 lbs seed
Cost of transplants				$6.49/128 = $0.06/plant
Transplanting labor				3 rows by hand: 3 hrs/2 beds total = $37.65L
				2 rows w/ transplanter, 6 beds at a time; 1 hr prep plants, 1.5hr x 3 people transplanting, 2 hrs machinery for 2 beds = $22.78L, $2.11 + .66 = $2.77M
Cultivation:				
Reemay on/off				For 2 beds: $105/3 uses = $35Pr, .75 hr laying = $9.41L
Hoeing 1x, 2x, 3x				at $12.55/hr: average 1 hr/2 beds — $12.55/2 beds
Hand weeding 1	100.40			at $12.55/hr: average 8 hrs/2 beds — $100.40/2 beds
Hand weeding 2				at $12.55/hr: average 4 hrs/2 beds — $50.20/2 beds
Hand weeding 3				at $12.55/hr: average 2 hrs/2 beds — $25.10/2 beds
Straw mulch				40 bales at $3, 1 hr/2 beds; $12.55L, $120.00Pr
Irrigating 1x	7.53	8.37		$7.53L, $8.37M per 2 beds, each use, w/ JD 2240
Tractor cultivating 6x	7.56	3.48		1A at a time: 1 hour/A = 6 mins/2 beds; $1.26L, $0.53 +.05 = $0.58M per pass w/ Cub mostly
Side-dressing				Spin 500 lbs 4-3-3/A, 1 hr total/20 beds = 6 mins/2 beds; $1.26L, $0.32 +.05 = $0.37M, $10Pr w/ Ford 4000
Spraying				1 hr/.5A total time = 12 mins/2 beds; $2.51L, $0.64 +.10 = $0.74M, $6Pr w/ Ford 4000
Flame weeding				10 beds/hr = 12 mins/2 beds; $2.51L, $0.64 +.10 = $0.74M, $6Pr w/ Ford 4000
Other				
Pre-harvest Subtotal:	134.34	16.50	75.00	= 225.84 Pre-harvest cost for two beds

Harvest:

Total yield for two 350' beds = 2000 bunches

Total hours to harvest two 350' beds: 13.3 hrs — at 150 bunches/hr

	Labor cost $	Machinery cost $	Product cost $	Notes
Field to pack house	166.92			at $12.55/hr — 13.3 hrs
Pack house to cooler	41.42			at $12.55/hr — at 600 bunches/hr
Bags, boxes, labels			89.88	$0.25/bag, $1.00/box, $0.07/label — 84 24-count boxes at $1.07
Delivery	30.12	9.60		See Worksheet 1.
Post Harvest:				
Mow crop				6 beds at a time: 10 mins/2 beds; $2.09L, $0.53 +.17 = $0.70M w/ Ford 4000
Remove mulch				1 hour/2 beds: $12.55L
Disk	1.26	0.73		$1.26L, $0.63 +.10 = $0.73M w/ JD 2240, see disking above.
Sow cover crop: spinner	1.26	0.68	8.00	1A at a time: 1 hr/20 beds = 6 mins/2 beds; $1.26L, $0.63 + .05 = $0.68M, $8Pr w/ JD 2240
Sow cover crop: Brillion				1A at a time: 2 hrs/20 beds = 12 mins/2 beds; $2.51L, $1.26 + .20 = $1.46M, 8Pr w/ JD 2240
Other				
Post-harvest Subtotal:	375.32	27.51	172.88	= 575.71 Harvested cost for 2 beds
Marketing Costs:				
Labor: sales calls for season (for this crop only)	6.28			Average 10 mins/week for 3 weeks: .5 hr
Commissions				Commissions, if any, to growers' co-op, broker, or salesperson
Farmers' market expense	60.24	4.70	9.00	See Worksheet 1.
Total Crop Costs:	441.84	32.21	181.88	= 655.93 Total crop costs
Overhead Costs:	288.00			Apportionment for two 350' beds, see Worksheet 1.
Total Costs: Crop & Overhead Total:	943.93			Total costs per two 350' beds

Sales:	# of units	Price per unit	Total $
Retail:	200.00	1.75	350.00
Wholesale:	1800.00	1.25	2250.00
Other:			0.00
Total units	2000.00		
Total Sales:			2600.00 For two 350' beds

Net Profit: Total sales – total costs =	1656.07	**Net profit for two 350' beds (1/10 acre)**
Net Profit/Acre:	16560.70	Standardize to one acre
Cost/Unit:	0.47	Total cost/total units
Net Profit/Unit:	0.83	Net profit/total units

NOTES:

Crop Enterprise Budget: Corn (sweet)

Crop Enterprise Budget

Crop Year: | Crop: **Corn: sweet** (and specify: early, mid, late) | Unit Area: **Two 350' beds**

Note: Twenty 350' beds = 1 acre

Bed feet or acres: **700' or 1/10A**

Today's Date: | Rows/bed&plant spacing: 2 rows/bed, direct seed, no mulch

Costs in $: Remember to prorate to unit area | Field:

NOTES: Labor at $12.55/hr. See Worksheet 1. Figures below are for two 350' beds.

	Labor cost $	Machinery cost $	Product cost $	Notes
Prepare Soil:				
Disk 1x	1.26	0.73		1A at a time: 1 hr total for 20 beds = 6 mins/2 beds; $1.26L, $0.63 + .10 = $0.73M w/ JD 2240; see Worksheet 4
Chisel 1x	2.51	0.74		.5A at a time:1 hr total for 10 beds = 12 mins/2 beds; $2.51L, $0.64 +.10 = $0.74M w/ Ford 4000; see Worksheet 4
Rototill 1x, 2x				.5A at a time: 2 hrs total for 10 beds = 24 mins/2 beds; $5.02L, $1.28 tractor + .52 tiller = $1.80M w/ Ford 4000
Bedform 2x	5.02	1.48		.5A at a time: 1 hr total for 10 beds = 12 mins/2 beds; $2.51L, $0.64 +.10 = $0.74M for ONE pass w/ Ford 4000
Fertilizer	1.26	0.68	10.00	500 lbs 4-3-3/A at a time: 1 hour total for 20 beds = 6 mins/2 beds; $.1.26L, $0.63 +.05 = $0.68M, $10Pr w/ JD 2240
Manure, compost	2.52	1.02	25.00	1A at a time: compost at $25/yd, 10 yds/A; 2 hrs total for 20 beds = 12 mins per 2 beds; $2.51L, $1.26 + .75 = $2.01M, $25Pr w/ JD 2240
Other				
Plastic mulch				.5A at a time: 1.5 hr/A laying = 10 mins/2 beds; $2.09L, $0.53 +.17 = $0.70M, $20Pr w/ Ford 4000
Seed/Transplant:				
Seeding in field	6.28		30.00	2 beds at a time: 30 mins/2 beds total = $6.28L 3000 seeds
Cost of transplants				$6.49/128 = $0.06/plant
Transplanting labor				3 rows by hand: 3 hrs/2 beds total = $37.65L
				2 rows w/ transplanter, 6 beds at a time; 1 hr prep plants, 1.5hr x 3 people transplanting, 2 hrs machinery for 2 beds = $22.78L, $2.11 + .66 = $2.77M
Cultivation:				
Reemay on/off				For 2 beds: $105/3 uses = $35Pr, .75 hr laying = $9.41L
Hoeing 1x, 2x, 3x	12.55			at $12.55/hr: average 1 hr/2 beds $12.55/2 beds
Hand weeding 1	50.20			at $12.55/hr: average 8 hrs/2 beds $100.40/2 beds
Hand weeding 2				at $12.55/hr: average 4 hrs/2 beds $50.20/2 beds
Hand weeding 3				at $12.55/hr: average 2 hrs/2 beds $25.10/2 beds
Straw mulch				40 bales at $3, 1 hr/2 beds; $12.55L, $120.00Pr
Irrigating 1x	7.53	8.37		$7.53L, $8.37M per 2 beds, each use, w/ JD 2240
Tractor cultivating 6x	7.56	3.48		1A at a time: 1 hour/A = 6 mins/2 beds; $1.26L, $0.53 +.05 = $0.58M per pass w/ Cub mostly
Side-dressing				Spin 500 lbs 4-3-3/A, 1 hr total/20 beds = 6 mins/2 beds; $1.26L, $0.32 +.05 = $0.37M, $10Pr w/ Ford 4000
Spraying	2.51	0.74	6.00	1 hr/.5A total time = 12 mins/2 beds; $2.51L, $0.64 +.10 = $0.74M, $6Pr w/ Ford 4000
Flame weeding				10 beds/hr = 12 mins/2 beds; $2.51L, $0.64 +.10 = $0.74M, $6Pr w/ Ford 4000
Other				
Pre-harvest Subtotal:	99.20	17.24	71.00	= 187.44 Pre-harvest cost for two beds

Harvest: Total yield for two 350' beds = 100 dozen (big ears only)
Total hours to harvest two 350' beds: 5 hrs (at 20 dozen/hr)

	Labor cost $	Machinery cost $	Product cost $	Notes
Field to pack house	62.75			at $12.55/hr 5 hrs
Pack house to cooler	25.10			at $12.55/hr at 50 dozen/hr
Bags, boxes, labels			34.25	$0.25/bag, $1.00/box, $0.07/label 25 boxes: 4 dozen/box, $1.37 per box
Delivery	30.12	9.60		See Worksheet 1.
Post Harvest:				
Mow crop	2.09	0.70		6 beds at a time: 10 mins/2 beds; $2.09L, $0.53 +.17 = $0.70M w/ Ford 4000
Remove mulch				1 hour/2 beds: $12.55L
Disk	1.26	0.73		$1.26L, $0.63 +.10 = $0.73M w/ JD 2240, see disking above.
Sow cover crop: spinner	1.26	0.68	8.00	1A at a time: 1 hr/20 beds = 6 mins/2 beds; $1.26L, $0.63 + .05 = $0.68M, $8Pr w/ JD 2240
Sow cover crop: Brillion				1A at a time: 2 hrs/20 beds = 12 mins/2 beds; $2.51L, $1.26 + .20 = $1.46M, 8Pr w/ JD 2240
Other				
Post-harvest Subtotal:	221.78	28.95	113.25	= 363.98 Harvested cost for 2 beds
Marketing Costs:				
Labor: sales calls for season (for this crop only)	6.28			Average 10 mins/week for 3 weeks: .5 hr
Commissions				Commissions, if any, to growers' co-op, broker, or salesperson
Farmers' market expense	60.24	4.70	9.00	See Worksheet 1.
Total Crop Costs:	288.30	33.65	122.25	= 444.20 Total crop costs
Overhead Costs:	288.00			Apportionment for two 350' beds, see Worksheet 1.
Total Costs: Crop & Overhead Total:	732.20			Total costs per two 350' beds

Sales:	# of units	Price per unit	Total $
Retail:	70.00	6.00	420.00
Wholesale:	30.00	4.00	120.00
Other:			0.00
Total units	100.00		
Total Sales:			540.00 For two 350' beds

Net Profit: Total sales – total costs =	-192.20	**Net profit for two 350' beds (1/10 acre)**
Net Profit/Acre:	-1922.00	Standardize to one acre
Cost/Unit:	7.32	Total cost/total units
Net Profit/Unit:	-1.92	Net profit/total units

NOTES:

Crop Enterprise Budget: Cucumbers

Crop Enterprise Budget

Crop year: | Crop: **Cucumbers** (and specify: early, mid, late) | **Unit Area:** **Two 350' beds** | **Note: Twenty 350' beds = 1 acre**

Bed feet or acres: **700' or 1/10A**

Today's Date: | Rows per bed & plant spacing: 1 row/bed, 3' hill spacing, plastic mulch

Costs in $: Remember to prorate to unit area | Field:

	Labor cost $	Machinery cost $	Product cost $	**NOTES:** Labor at $12.55/hr. See Worksheet 1. Figures below are for two 350' beds.
Prepare Soil:				
Disk 1x	1.26	0.73		1A at a time: 1 hr total for 20 beds = 6 mins/2 beds; $1.26L, $0.63 + .10 = $0.73M w/ JD 2240; see Worksheet 4
Chisel 1x	2.51	0.74		.5A at a time:1 hr total for 10 beds = 12 mins/2 beds; $2.51L, $0.64 +.10 = $0.74M w/ Ford 4000; see Worksheet 4
Rototill 1x, 2x				.5A at a time: 2 hrs total for 10 beds = 24 mins/2 beds; $5.02L, $1.28 tractor + .52 tiller = $1.80M w/ Ford 4000
Bedform 2x	5.02	1.48		.5A at a time: 1 hr total for 10 beds = 12 mins/2 beds; $2.51L, $0.64 +.10 = $0.74M for ONE pass w/ Ford 4000
Fertilizer	1.26	0.68	10.00	500 lbs 4-3-3/A at a time: 1 hour total for 20 beds = 6 mins/2 beds; $.1.26L, $0.63 +.05 = $0.68M, $10Pr w/ JD 2240
Manure, compost	2.52	1.02	25.00	1A at a time: compost at $25/yd, 10 yds/A; 2 hrs total for 20 beds = 12 mins per 2 beds; $2.51L, $1.26 + .75 = $2.01M, $25Pr w/ JD 2240
Other				
Plastic mulch	2.09	0.70	20.00	.5A at a time: 1.5 hr/A laying = 10 mins/2 beds; $2.09L, $0.53 +.17 = $0.70M, $20Pr w/ Ford 4000
Seed/Transplant:				
Seeding in field				2 beds at a time: 30 mins/2 beds total = $6.28L
Cost of transplants			47.00	$6.49/128 = $0.06/plant; 235 plants in 804s: $0.20/plant
Transplanting labor	25.10			3 rows by hand: 3 hrs/2 beds total = $37.65L; 2 hrs total
				2 rows w/ transplanter, 6 beds at a time; 1 hr prep plants, 1.5hr x 3 people transplanting, 2 hrs machinery for 2 beds = $22.78L, $2.11 + .66 = $2.77M
Cultivation:				
Reemay on/off	9.41		35.00	For 2 beds: $105/3 uses = $35Pr, .75 hr laying = $9.41L
Hoeing 1x, 2x, 3x	25.10			at $12.55/hr: average 1 hr/2 beds; $12.55/2 beds; hoe edges of mulch 2x
Hand weeding 1	12.55			at $12.55/hr: average 8 hrs/2 beds; $100.40/2 beds
Hand weeding 2	12.55			at $12.55/hr: average 4 hrs/2 beds; $50.20/2 beds
Hand weeding 3				at $12.55/hr: average 2 hrs/2 beds; $25.10/2 beds
Straw mulch				40 bales at $3, 1 hr/2 beds; $12.55L, $120.00Pr
Irrigating 1x	7.53	8.37		$7.53L, $8.37M per 2 beds, each use, w/ JD 2240
Tractor cultivating 6x	7.56	3.48		1A at a time: 1 hour/A = 6 mins/2 beds; $1.26L, $0.53 +.05 = $0.58M per pass w/ Cub mostly
Side-dressing				Spin 500 lbs 4-3-3/A, 1 hr total/20 beds = 6 mins/2 beds; $1.26L, $0.32 +.05 = $0.37M, $10Pr w/ Ford 4000
Spraying				1 hr/.5A total time = 12 mins/2 beds; $2.51L, $0.64 +.10 = $0.74M, $6Pr w/ Ford 4000
Flame weeding				10 beds/hr = 12 mins/2 beds; $2.51L, $0.64 +.10 = $0.74M, $6Pr w/ Ford 4000
Other				
Pre-harvest Subtotal:	114.46	17.20	137.00	= 268.66 Pre-harvest cost for two beds

Harvest:

Total yield for two 350' beds = 25 50-lb boxes

Total hours to harvest two 350' beds 8.3 hrs at 3 cases/hr

	Labor cost $	Machinery cost $	Product cost $	Notes
Field to pack house	37.65			at $12.55/hr; 8.3 hrs
Pack house to cooler	31.38			at $12.55/hr; at 10 cases/hr
Bags, boxes, labels			26.75	$0.25/bag, $1.00/box, $0.07/label; 25 boxes at $1.07
Delivery	30.12	9.60		See Worksheet 1.
Post Harvest:				
Mow crop				6 beds at a time: 10 mins/2 beds; $2.09L, $0.53 +.17 = $0.70M w/ Ford 4000
Remove mulch	12.55			1 hour/2 beds: $12.55L
Disk	1.26	0.73		$1.26L, $0.63 +.10 = $0.73M w/ JD 2240, see disking above.
Sow cover crop: spinner	1.26	0.68	8.00	1A at a time: 1 hr/20 beds = 6 mins/2 beds; $1.26L, $0.63 + .05 = $0.68M, $8Pr w/ JD 2240
Sow cover crop: Brillion				1A at a time: 2 hrs/20 beds = 12 mins/2 beds; $2.51L, $1.26 + .20 = $1.46M, 8Pr w/ JD 2240
Other				
Post-harvest Subtotal:	228.68	28.21	171.75	= 428.64 Harvested cost for 2 beds
Marketing Costs:				
Labor: sales calls for season (for this crop only)	6.28			Average 10 mins/week for 3 weeks: .5 hr
Commissions				Commissions, if any, to growers' co-op, broker, or salesperson
Farmers' market expense	60.24	4.70	9.00	See Worksheet 1.
Total Crop Costs:	295.20	32.91	180.75	= 508.86 Total crop costs
Overhead Costs:	288.00			Apportionment for two 350' beds, see Worksheet 1.
Total Costs: **Crop & Overhead Total:**	796.86			Total costs per two 350' beds

NOTES:

Sales:	# of units	Price per unit	Total $
Retail:	10.00	50.00	500.00
Wholesale:	15.00	30.00	450.00
Other:			0.00
Total units	25.00		
Total Sales:			950.00 For two 350' beds

Net Profit: Total sales – total costs =	153.14	**Net profit for two 350' beds (1/10 acre)**
Net Profit/Acre:	1531.40	Standardize to one acre
Cost/Unit:	31.87	Total cost/total units
Net Profit/Unit:	6.13	Net profit/total units

Crop Enterprise Budget: Dill (bunches)

Crop Enterprise Budget

Crop Year: | Crop: **Dill: bunches** (and specify: early, mid, late) | **Unit Area:** **Two 350' beds** | Note: Twenty 350' beds = 1 acre

Bed feet or acres: **700' or 1/10A**

Today's Date: | Rows per bed & plant spacing: 3 rows/bed, thickly seeded

Costs in $: Remember to prorate to unit area | Field:

NOTES: Labor at $12.55/hr. See Worksheet 1. Figures below are for two 350' beds.

	Labor cost $	Machinery cost $	Product cost $	Notes
Prepare Soil:				
Disk 1x	1.26	0.73		1A at a time: 1 hr total for 20 beds = 6 mins/2 beds; $1.26L, $0.63 + .10 = $0.73M w/ JD 2240; see Worksheet 4
Chisel 1x	2.51	0.74		.5A at a time:1 hr total for 10 beds = 12 mins/2 beds; $2.51L, $0.64 +.10 = $0.74M w/ Ford 4000; see Worksheet 4
Rototill 1x, 2x				.5A at a time: 2 hrs total for 10 beds = 24 mins/2 beds; $5.02L, $1.28 tractor + .52 tiller = $1.80M w/ Ford 4000
Bedform 2x	5.02	1.48		.5A at a time: 1 hr total for 10 beds = 12 mins/2 beds; $2.51L, $0.64 +.10 = $0.74M for ONE pass w/ Ford 4000
Fertilizer	1.26	0.68	10.00	500 lbs 4-3-3/A at a time: 1 hour total for 20 beds = 6 mins/2 beds; $.1.26L, $0.63 +.05 = $0.68M, $10Pr w/ JD 2240
Manure, compost	2.52	1.02	25.00	1A at a time: compost at $25/yd, 10 yds/A; 2 hrs total for 20 beds = 12 mins per 2 beds; $2.51L, $1.26 + .75 = $2.01M, $25Pr w/ JD 2240
Other				
Plastic mulch				.5A at a time: 1.5 hr/A laying = 10 mins/2 beds; $2.09L, $0.53 +.17 = $0.70M, $20Pr w/ Ford 4000
Seed/Transplant:				
Seeding in field	6.28		39.00	2 beds at a time: 30 mins/2 beds total = $6.28L — 2 lbs seed
Cost of transplants				$6.49/128 = $0.06/plant
Transplanting labor				3 rows by hand: 3 hrs/2 beds total = $37.65L
				2 rows w/ transplanter, 6 beds at a time; 1 hr prep plants, 1.5hr x 3 people transplanting, 2 hrs machinery for 2 beds = $22.78L, $2.11 + .66 = $2.77
Cultivation:				
Reemay on/off				For 2 beds: $105/3 uses = $35Pr, .75 hr laying = $9.41L
Hoeing 1x, 2x, 3x				at $12.55/hr: average 1 hr/2 beds — $12.55/2 beds
Hand weeding 1	100.40			at $12.55/hr: average 8 hrs/2 beds — $100.40/2 beds
Hand weeding 2				at $12.55/hr: average 4 hrs/2 beds — $50.20/2 beds
Hand weeding 3				at $12.55/hr: average 2 hrs/2 beds — $25.10/2 beds
Straw mulch				40 bales at $3, 1 hr/2 beds; $12.55L, $120.00Pr
Irrigating 1x	7.53	8.37		$7.53L, $8.37M per 2 beds, each use, w/ JD 2240
Tractor cultivating 6x	7.56	3.48		1A at a time: 1 hour/A = 6 mins/2 beds; $1.26L, $0.53 +.05 = $0.58M per pass w/ Cub mostly
Side-dressing				Spin 500 lbs 4-3-3/A, 1 hr total/20 beds = 6 mins/2 beds; $1.26L, $0.32 +.05 = $0.37M, $10Pr w/ Ford 4000
Spraying				1 hr/.5A total time = 12 mins/2 beds; $2.51L, $0.64 +.10 = $0.74M, $6Pr w/ Ford 4000
Flame weeding				10 beds/hr = 12 mins/2 beds; $2.51L, $0.64 +.10 = $0.74M, $6Pr w/ Ford 4000
Other				
Pre-harvest Subtotal:	134.34	16.50	74.00	= 224.84 Pre-harvest cost for two beds

Harvest:

Total yield for two 350' beds = 2000 bunches

Total hours to harvest two 350' beds 16 hrs at 125 bunches/hr

	Labor cost $	Machinery cost $	Product cost $	Notes
Field to pack house	200.80			at $12.55/hr — 16 hrs
Pack house to cooler	41.42			at $12.55/hr — at 600 bunches/hr
Bags, boxes, labels			89.88	$0.25/bag, $1.00/box, $0.07/label — 84 at $1.07
Delivery	30.12	9.60		See Worksheet 1.
Post Harvest:				
Mow crop				6 beds at a time: 10 mins/2 beds; $2.09L, $0.53 +.17 = $0.70M w/ Ford 4000
Remove mulch				1 hour/2 beds: $12.55L
Disk	1.26	0.73		$1.26L, $0.63 +.10 = $0.73M w/ JD 2240, see disking above.
Sow cover crop: spinner	1.26	0.68	8.00	1A at a time: 1 hr/20 beds = 6 mins/2 beds; $1.26L, $0.63 + .05 = $0.68M, $8Pr w/ JD 2240
Sow cover crop: Brillion				1A at a time: 2 hrs/20 beds = 12 mins/2 beds; $2.51L, $1.26 + .20 = $1.46M, 8Pr w/ JD 2240
Other				
Post-harvest Subtotal:	409.20	27.51	171.88	= 608.59 Harvested cost for 2 beds
Marketing Costs:				
Labor: sales calls for season (for this crop only)	6.28			Average 10 mins/week for 3 weeks: .5 hr
Commissions				Commissions, if any, to growers' co-op, broker, or salesperson
Farmers' market expense	60.24	4.70	9.00	See Worksheet 1.
Total Crop Costs:	475.72	32.21	180.88	= 688.81 Total crop costs
Overhead Costs:	288.00			Apportionment for two 350' beds, see Worksheet 1.
Total Costs: Crop & Overhead Total:	976.81			Total costs per two 350' beds

Sales:	# of units	Price per unit	Total $
Retail:	200.00	1.75	350.00
Wholesale:	1800.00	1.25	2250.00
Other:			0.00
Total units	2000.00		
Total Sales:			2600.00 For two 350' beds

Net Profit: Total sales – total costs =	1623.19	**Net profit for two 350' beds (1/10 acre)**
Net Profit/Acre:	16231.90	Standardize to one acre
Cost/Unit:	0.49	Total cost/total units
Net Profit/Unit:	0.81	Net profit/total units

NOTES:

Crop Enterprise Budget: Kale (bunches)

Crop Enterprise Budget

Crop Year: | Crop: **Kale: bunches** (and specify: early, mid, late) | **Unit Area: Two 350' beds** | **Note: Twenty 350' beds = 1 acre**

Bed feet or acres: **700' or 1/10A**

Today's Date: | Rows per bed & plant spacing: 2 rows/bed, 24" spacing, transplanted

Costs in $: Remember to prorate to unit area | Field:

NOTES: Labor at $12.55/hr. See Worksheet 1. Figures below are for two 350' beds.

Prepare Soil:	Labor cost $	Machinery cost $	Product cost $	Notes
Disk 1x	1.26	0.73		1A at a time: 1 hr total for 20 beds = 6 mins/2 beds; $1.26L, $0.63 + .10 = $0.73M w/ JD 2240; see Worksheet 4
Chisel 1x	2.51	0.74		.5A at a time:1 hr total for 10 beds = 12 mins/2 beds; $2.51L, $0.64 +.10 = $0.74M w/ Ford 4000; see Worksheet 4
Rototill 1x, 2x				.5A at a time: 2 hrs total for 10 beds = 24 mins/2 beds; $5.02L, $1.28 tractor + .52 tiller = $1.80M w/ Ford 4000
Bedform 2x	5.02	1.48		.5A at a time: 1 hr total for 10 beds = 12 mins/2 beds; $2.51L, $0.64 +.10 = $0.74M for ONE pass w/ Ford 4000
Fertilizer	1.26	0.68	10.00	500 lbs 4-3-3/A at a time: 1 hour total for 20 beds = 6 mins/2 beds; $.1.26L, $0.63 +.05 = $0.68M, $10Pr w/ JD 2240
Manure, compost	2.52	1.02	25.00	1A at a time: compost at $25/yd, 10 yds/A; 2 hrs total for 20 beds = 12 mins per 2 beds; $2.51L, $1.26 + .75 = $2.01M, $25Pr w/ JD 2240
Other				
Plastic mulch				.5A at a time: 1.5 hr/A laying = 10 mins/2 beds; $2.09L, $0.53 +.17 = $0.70M, $20Pr w/ Ford 4000

Seed/Transplant:	Labor cost	Machinery cost	Product cost	Notes
Seeding in field				2 beds at a time: 30 mins/2 beds total = $6.28L
Cost of transplants			42.00	$6.49/128 = $0.06/plant — 700 plants
Transplanting labor	25.23			3 rows by hand: 3 hrs/2 beds total = $37.65L — 2/3 of 3-row time
				2 rows w/ transplanter, 6 beds at a time; 1 hr prep plants, 1.5hr x 3 people transplanting, 2 hrs machinery for 2 beds = $22.78L, $2.11 + .66 = $2.77M

Cultivation:	Labor cost	Machinery cost	Product cost	Notes
Reemay on/off				For 2 beds: $105/3 uses = $35Pr, .75 hr laying = $9.41L
Hoeing 1x, 2x, 3x	25.10			at $12.55/hr: average 1 hr/2 beds — $12.55/2 beds
Hand weeding 1	50.20			at $12.55/hr: average 8 hrs/2 beds — $100.40/2 beds
Hand weeding 2	25.10			at $12.55/hr: average 4 hrs/2 beds — $50.20/2 beds
Hand weeding 3				at $12.55/hr: average 2 hrs/2 beds — $25.10/2 beds
Straw mulch				40 bales at $3, 1 hr/2 beds; $12.55L, $120.00Pr
Irrigating 1x	7.53	8.37		$7.53L, $8.37M per 2 beds, each use, w/ JD 2240
Tractor cultivating 6x	7.56	3.48		1A at a time: 1 hour/A = 6 mins/2 beds; $1.26L, $0.53 +.05 = $0.58M per pass w/ Cub mostly
Side-dressing				Spin 500 lbs 4-3-3/A, 1 hr total/20 beds = 6 mins/2 beds; $1.26L, $0.32 +.05 = $0.37M, $10Pr w/ Ford 4000
Spraying	5.02	1.48	12.00	1 hr/.5A total time = 12 mins/2 beds; $2.51L, $0.64 +.10 = $0.74M, $6Pr w/ Ford 4000
Flame weeding				10 beds/hr = 12 mins/2 beds; $2.51L, $0.64 +.10 = $0.74M, $6Pr w/ Ford 4000
Other				
Pre-harvest Subtotal:	158.31	17.98	89.00	= 265.29 Pre-harvest cost for two beds

Harvest:

Total yield for two 350' beds = 2800 bunches

Total hours to harvest two 350' beds 18.7 hrs — 150 bunches/hr

	Labor cost	Machinery cost	Product cost	Notes
Field to pack house	234.69			at $12.55/hr — 18.7 hrs
Pack house to cooler	292.42			at $12.55/hr — 120 bunches/hr, 23.3 hrs
Bags, boxes, labels			166.92	$0.25/bag, $1.00/box, $0.07/label — 156 18-count boxes at $1.07
Delivery	30.12	9.60		See Worksheet 1.

Post Harvest:	Labor cost	Machinery cost	Product cost	Notes
Mow crop	2.09	0.70		6 beds at a time: 10 mins/2 beds; $2.09L, $0.53 +.17 = $0.70M w/ Ford 4000
Remove mulch				1 hour/2 beds: $12.55L
Disk	1.26	0.73		$1.26L, $0.63 +.10 = $0.73M w/ JD 2240, see disking above.
Sow cover crop: spinner	1.26	0.68	8.00	1A at a time: 1 hr/20 beds = 6 mins/2 beds; $1.26L, $0.63 + .05 = $0.68M, $8Pr w/ JD 2240
Sow cover crop: Brillion				1A at a time: 2 hrs/20 beds = 12 mins/2 beds; $2.51L, $1.26 + .20 = $1.46M, 8Pr w/ JD 2240
Other				
Post-harvest Subtotal:	720.15	29.69	263.92	= 1013.76 Harvested cost for 2 beds

Marketing Costs:	Labor cost	Machinery cost	Product cost	Notes
Labor: sales calls for season (for this crop only)	6.28			Average 10 mins/week for 3 weeks: .5 hr
Commissions				Commissions, if any, to growers' co-op, broker, or salesperson
Farmers' market expense	60.24	4.70	9.00	See Worksheet 1.
Total Crop Costs:	786.67	34.39	272.92	= 1093.98 Total crop costs
Overhead Costs:	288.00			Apportionment for two 350' beds, see Worksheet 1.
Total Costs: Crop & Overhead Total:	1381.98			Total costs per two 350' beds

Sales:	# of units	Price per unit	Total $
Retail:	460.00	2.00	920.00
Wholesale:	2340.00	1.25	2925.00
Other:			0.00
Total units	2800.00		
Total Sales:			3845.00 For two 350' beds

Net Profit: Total sales – total costs =	2463.02	**Net profit for two 350' beds (1/10 acre)**
Net Profit/Acre:	24630.20	Standardize to one acre
Cost/Unit:	0.49	Total cost/total units
Net Profit/Unit:	0.88	Net profit/total units

NOTES:

Crop Enterprise Budget: Lettuce

Crop Enterprise Budget

Crop Year:

Crop: **Lettuce: heads** (and specify: early, mid, late)

Unit Area: **Two 350' beds**

Bed feet or acres: **700' or 1/10A**

Note: Twenty 350' beds = 1 acre

Today's Date:

Rows per bed & plant spacing: 3 rows/bed, 16' spacing, transplanted

Costs in $: Remember to prorate to unit area

Field:

NOTES: Labor at $12.55/hr. See Worksheet 1. Figures below are for two 350' beds.

	Labor cost $	Machinery cost $	Product cost $	Notes
Prepare Soil:				
Disk 1x	1.26	0.73		1A at a time: 1 hr total for 20 beds = 6 mins/2 beds; $1.26L, $0.63 + .10 = $0.73M w/ JD 2240; see Worksheet 4
Chisel 1x	2.51	0.74		.5A at a time:1 hr total for 10 beds = 12 mins/2 beds; $2.51L, $0.64 +.10 = $0.74M w/ Ford 4000; see Worksheet 4
Rototill 1x, 2x				.5A at a time: 2 hrs total for 10 beds = 24 mins/2 beds; $5.02L, $1.28 tractor + .52 tiller = $1.80M w/ Ford 4000
Bedform 2x	5.02	1.48		.5A at a time: 1 hr total for 10 beds = 12 mins/2 beds; $2.51L, $0.64 +.10 = $0.74M for ONE pass w/ Ford 4000
Fertilizer	1.26	0.68	10.00	500 lbs 4-3-3/A at a time: 1 hour total for 20 beds = 6 mins/2 beds; $.1.26L, $0.63 +.05 = $0.68M, $10Pr w/ JD 2240
Manure, compost	2.52	1.02	25.00	1A at a time: compost at $25/yd, 10 yds/A; 2 hrs total for 20 beds = 12 mins per 2 beds; $2.51L, $1.26 + .75 = $2.01M, $25Pr w/ JD 2240
Other				
Plastic mulch				.5A at a time: 1.5 hr/A laying = 10 mins/2 beds; $2.09L, $0.53 +.17 = $0.70M, $20Pr w/ Ford 4000
Seed/Transplant:				
Seeding in field				2 beds at a time: 30 mins/2 beds total = $6.28L
Cost of transplants			94.50	$6.49/128 = $0.06/plant 1575 plants
Transplanting labor	37.65			3 rows by hand: 3 hrs/2 beds total = $37.65L 2 rows w/ transplanter, 6 beds at a time; 1 hr prep plants, 1.5hr x 3 people transplanting, 2 hrs machinery for 2 beds = $22.78L, $2.11 + .66 = $2.77M
Cultivation:				
Reemay on/off				For 2 beds: $105/3 uses = $35Pr, .75 hr laying = $9.41L
Hoeing 1x, 2x, 3x	25.10			at $12.55/hr: average 1 hr/2 beds $12.55/2 beds
Hand weeding 1	25.10			at $12.55/hr: average 8 hrs/2 beds $100.40/2 beds
Hand weeding 2				at $12.55/hr: average 4 hrs/2 beds $50.20/2 beds
Hand weeding 3				at $12.55/hr: average 2 hrs/2 beds $25.10/2 beds
Straw mulch				40 bales at $3, 1 hr/2 beds; $12.55L, $120.00Pr
Irrigating 1x	7.53	8.37		$7.53L, $8.37M per 2 beds, each use, w/ JD 2240
Tractor cultivating 6x	7.56	3.48		1A at a time: 1 hour/A = 6 mins/2 beds; $1.26L, $0.53 +.05 = $0.58M per pass w/ Cub mostly
Side-dressing				Spin 500 lbs 4-3-3/A, 1 hr total/20 beds = 6 mins/2 beds; $1.26L, $0.32 +.05 = $0.37M, $10Pr w/ Ford 4000
Spraying				1 hr/.5A total time = 12 mins/2 beds; $2.51L, $0.64 +.10 = $0.74M, $6Pr w/ Ford 4000
Flame weeding				10 beds/hr = 12 mins/2 beds; $2.51L, $0.64 +.10 = $0.74M, $6Pr w/ Ford 4000
Other				
Pre-harvest Subtotal:	115.51	16.50	129.50	= 261.51 Pre-harvest cost for two beds

Harvest:

Total yield for two 350' beds = 52 24-ct cases 1248 marketable heads

Total hours to harvest two 350' beds 10.4 hrs 5 cases/hr

	Labor cost $	Machinery cost $	Product cost $	Notes
Field to pack house	130.52			at $12.55/hr 10.4 hrs
Pack house to cooler	130.52			at $12.55/hr at 5 cases/hr
Bags, boxes, labels			79.04	$0.25/bag, $1.00/box, $0.07/label 52 boxes at $1.52
Delivery	30.12	9.60		See Worksheet 1.
Post Harvest:				
Mow crop				6 beds at a time: 10 mins/2 beds; $2.09L, $0.53 +.17 = $0.70M w/ Ford 4000
Remove mulch				1 hour/2 beds: $12.55L
Disk	1.26	0.73		$1.26L, $0.63 +.10 = $0.73M w/ JD 2240, see disking above.
Sow cover crop: spinner	1.26	0.68	8.00	1A at a time: 1 hr/20 beds = 6 mins/2 beds; $1.26L, $0.63 + .05 = $0.68M, $8Pr w/ JD 2240
Sow cover crop: Brillion				1A at a time: 2 hrs/20 beds = 12 mins/2 beds; $2.51L, $1.26 + .20 = $1.46M, 8Pr w/ JD 2240
Other				
Post-harvest Subtotal:	409.19	27.51	216.54	= 653.24 Harvested cost for 2 beds
Marketing Costs:				
Labor: sales calls for season (for this crop only)	6.28			Average 10 mins/week for 3 weeks: .5 hr
Commissions				Commissions, if any, to growers' co-op, broker, or salesperson
Farmers' market expense	60.24	4.70	9.00	See Worksheet 1.
Total Crop Costs:	475.71	32.21	225.54	= 733.46 Total crop costs
Overhead Costs:	288.00			Apportionment for two 350' beds, see Worksheet 1.
Total Costs: Crop & Overhead Total:	1021.46			Total costs per two 350' beds

Sales:

	# of units	Price per unit	Total $
Retail:	16.00	48.00	768.00
Wholesale:	36.00	29.00	1044.00
Other:			0.00
Total units	52.00		
Total Sales:			1812.00 For two 350' beds

Net Profit: Total sales – total costs =	790.54	**Net profit for two 350' beds (1/10 acre)**
Net Profit/Acre:	7905.40	Standardize to one acre
Cost/Unit:	19.64	Total cost/total units
Net Profit/Unit:	15.20	Net profit/total units

NOTES:

Crop Enterprise Budget: Onions

Crop Enterprise Budget

Crop year:		Crop:	Onions	
		and specify: early, mid, late		
Unit Area:	Two 350' beds			Note: Twenty 350' beds = 1 acre
Bed feet or acres:	700' or 1/10A			
Today's Date:		Rows per bed & plant spacing:	3 rows/bed, 6" in row, 2 plants/each hole, on plastic mulch	
Costs in $:	Remember to prorate to unit area	Field:		

NOTES: Labor at $12.55/hr. See Worksheet 1. Figures below are for two 350' beds

	Labor cost $	Machinery cost $	Product cost $	Notes
Prepare Soil:				
Disk 1x	1.26	0.73		1A at a time: 1 hr total for 20 beds = 6 mins/2 beds; $1.26L, $0.63 + .10 = $0.73M w/ JD 2240; see Worksheet 4
Chisel 1x	2.51	0.74		.5A at a time:1 hr total for 10 beds = 12 mins/2 beds; $2.51L, $0.64 +.10 = $0.74M w/ Ford 4000; see Worksheet 4
Rototill 1x, 2x				.5A at a time: 2 hrs total for 10 beds = 24 mins/2 beds; $5.02L, $1.28 tractor + .52 tiller = $1.80M w/ Ford 4000
Bedform 2x	5.02	1.48		.5A at a time: 1 hr total for 10 beds = 12 mins/2 beds; $2.51L, $0.64 +.10 = $0.74M for ONE pass w/ Ford 4000
Fertilizer	1.26	0.68	10.00	500 lbs 4-3-3/A at a time: 1 hour total for 20 beds = 6 mins/2 beds; $.1.26L, $0.63 +.05 = $0.68M, $10Pr w/ JD 2240
Manure, compost	2.52	1.02	25.00	1A at a time: compost at $25/yd, 10 yds/A; 2 hrs total for 20 beds = 12 mins per 2 beds; $2.51L, $1.26 + .75 = $2.01M, $25Pr w/ JD 2240
Other				
Plastic mulch	2.09	0.70	20.00	.5A at a time: 1.5 hr/A laying = 10 mins/2 beds; $2.09L, $0.53 +.17 = $0.70M, $20Pr w/ Ford 4000
Seed/Transplant:				
Seeding in field				2 beds at a time: 30 mins/2 beds total = $6.28L
Cost of transplants			143.43	$6.49/128 = $0.06/plant; 8400 plants: 400 pl/open 1020, 21 1020 trays, at $6.83
Transplanting labor	112.95			3 rows by hand: 3 hrs/2 beds total = $37.65L; slower rate through plastic
				2 rows w/ transplanter, 6 beds at a time; 1 hr prep plants, 1.5hr x 3 people transplanting, 2 hrs machinery for 2 beds = $22.78L, $2.11 + .66 = $2.77M
Cultivation:				
Reemay on/off				For 2 beds: $105/3 uses = $35Pr, .75 hr laying = $9.41L
Hoeing 1x, 2x, 3x	25.10			at $12.55/hr: average 1 hr/2 beds; $12.55/2 beds; hoe edges of plastic
Hand weeding 1	25.10			at $12.55/hr: average 8 hrs/2 beds; $100.40/2 beds; weed holes of plastic
Hand weeding 2	12.55			at $12.55/hr: average 4 hrs/2 beds; $50.20/2 beds
Hand weeding 3	12.55			at $12.55/hr: average 2 hrs/2 beds; $25.10/2 beds
Straw mulch				40 bales at $3, 1 hr/2 beds; $12.55L, $120.00Pr
Irrigating 1x	7.53	8.37		$7.53L, $8.37M per 2 beds, each use, w/ JD 2240
Tractor cultivating 6x	7.56	3.48		1A at a time: 1 hour/A = 6 mins/2 beds; $1.26L, $0.53 +.05 = $0.58M per pass w/ Cub mostly
Side-dressing				Spin 500 lbs 4-3-3/A, 1 hr total/20 beds = 6 mins/2 beds; $1.26L, $0.32 +.05 = $0.37M, $10Pr w/ Ford 4000
Spraying				1 hr/.5A total time = 12 mins/2 beds; $2.51L, $0.64 +.10 = $0.74M, $6Pr w/ Ford 4000
Flame weeding				10 beds/hr = 12 mins/2 beds; $2.51L, $0.64 +.10 = $0.74M, $6Pr w/ Ford 4000
Other	37.65			
Pre-harvest Subtotal:	255.65	17.20	198.43	= 471.28 Pre-harvest cost for two beds

Harvest:

Total yield for two 350' beds = 40 50-lb bags

Total hours to harvest two 350' beds 2 hrs — 20 bags/hr

	Labor cost $	Machinery cost $	Product cost $	Notes
Field to pack house	25.10			at $12.55/hr; 2 hrs
Pack house to cooler	50.20			at $12.55/hr; at 10 bags/hr
Bags, boxes, labels			10.00	$0.25/bag, $1.00/box, $0.07/label; 40 at $0.25
Delivery	30.12	9.60		See Worksheet 1.
Post Harvest:				
Mow crop				6 beds at a time: 10 mins/2 beds; $2.09L, $0.53 +.17 = $0.70M w/ Ford 4000
Remove mulch	12.55			1 hour/2 beds: $12.55L
Disk	1.26	0.73		$1.26L, $0.63 +.10 = $0.73M w/ JD 2240, see disking above.
Sow cover crop: spinner	1.26	0.68	8.00	1A at a time: 1 hr/20 beds = 6 mins/2 beds; $1.26L, $0.63 + .05 = $0.68M, $8Pr w/ JD 2240
Sow cover crop: Brillion				1A at a time: 2 hrs/20 beds = 12 mins/2 beds; $2.51L, $1.26 + .20 = $1.46M, 8Pr w/ JD 2240
Other				
Post-harvest Subtotal:	376.14	28.21	216.43	= 620.78 Harvested cost for 2 beds
Marketing Costs:				
Labor: sales calls for season (for this crop only)	6.28			Average 10 mins/week for 3 weeks: .5 hr
Commissions				Commissions, if any, to growers' co-op, broker, or salesperson
Farmers' market expense	60.24	4.70	9.00	See Worksheet 1.
Total Crop Costs:	442.66	32.91	225.43	= 701.00 Total crop costs
Overhead Costs:	288.00			Apportionment for two 350' beds, see Worksheet 1.
<u>Total Costs:</u> **Crop & Overhead Total:**	989.00			Total costs per two 350' beds

Sales:	# of units	Price per unit	Total $
Retail:	20.00	50.00	1000.00
Wholesale:	20.00	30.00	600.00
Other:			0.00
Total units	40.00		
<u>Total Sales:</u>			1600.00 For two 350' beds

Net Profit: Total sales – total costs =	611.00	**Net profit for two 350' beds (1/10 acre)**
Net Profit/Acre:	6110.00	Standardize to one acre
Cost/Unit:	24.73	Total cost/total units
<u>Net Profit/Unit:</u>	15.28	Net profit/total units

NOTES:

Crop Enterprise Budget: Parsley (bunches)

Crop Enterprise Budget

Crop Year:	
Crop: (and specify: early, mid, late)	Parsley: bunches
Unit Area:	Two 350' beds
Bed feet or acres:	700' or 1/10A
Today's Date:	
Rows per bed & plant spacing:	3 rows/bed, 6" apart, transplanted on plastic
Field:	

Note: Twenty 350' beds = 1 acre

Costs in $: Remember to prorate to unit area

NOTES: Labor at $12.55/hr. See Worksheet 1. Figures below are for two 350' beds

	$ Labor cost	$ Machinery cost	$ Product cost	Notes		
Prepare Soil:						
Disk 1x	1.26	0.73		1A at a time: 1 hr total for 20 beds = 6 mins/2 beds; $1.26L, $0.63 + .10 = $0.73M w/ JD 2240; see Worksheet 4		
Chisel 1x	2.51	0.74		.5A at a time:1 hr total for 10 beds = 12 mins/2 beds; $2.51L, $0.64 +.10 = $0.74M w/ Ford 4000; see Worksheet 4		
Rototill 1x, 2x				.5A at a time: 2 hrs total for 10 beds = 24 mins/2 beds; $5.02L, $1.28 tractor + .52 tiller = $1.80M w/ Ford 4000		
Bedform 2x	5.02	1.48		.5A at a time: 1 hr total for 10 beds = 12 mins/2 beds; $2.51L, $0.64 +.10 = $0.74M for ONE pass w/ Ford 4000		
Fertilizer	1.26	0.68	10.00	500 lbs 4-3-3/A at a time: 1 hour total for 20 beds = 6 mins/2 beds; $.1.26L, $0.63 +.05 = $0.68M, $10Pr w/ JD 2240		
Manure, compost	2.52	1.02	25.00	1A at a time: compost at $25/yd, 10 yds/A; 2 hrs total for 20 beds = 12 mins per 2 beds; $2.51L, $1.26 + .75 = $2.01M, $25Pr w/ JD 2240		
Other						
Plastic mulch	2.09	0.70	20.00	.5A at a time: 1.5 hr/A laying = 10 mins/2 beds; $2.09L, $0.53 +.17 = $0.70M, $20Pr w/ Ford 4000		
Seed/Transplant:						
Seeding in field				2 beds at a time: 30 mins/2 beds total = $6.28L		
Cost of transplants			252.00	$6.49/128 = $0.06/plant	4200 plants at $0.06	
Transplanting labor	112.95			3 rows by hand: 3 hrs/2 beds total = $37.65L	slower planting through plastic	
				2 rows w/ transplanter, 6 beds at a time; 1 hr prep plants, 1.5hr x 3 people transplanting, 2 hrs machinery for 2 beds = $22.78L, $2.11 + .66 = $2.77M		
Cultivation:						
Reemay on/off				For 2 beds: $105/3 uses = $35Pr, .75 hr laying = $9.41L		
Hoeing 1x, 2x, 3x	25.10			at $12.55/hr: average 1 hr/2 beds	$12.55/2 beds	hoe edges of plastic
Hand weeding 1	25.10			at $12.55/hr: average 8 hrs/2 beds	$100.40/2 beds	weeding holes, edges
Hand weeding 2	12.55			at $12.55/hr: average 4 hrs/2 beds	$50.20/2 beds	
Hand weeding 3	12.55			at $12.55/hr: average 2 hrs/2 beds	$25.10/2 beds	
Straw mulch				40 bales at $3, 1 hr/2 beds; $12.55L, $120.00Pr		
Irrigating 1x	7.53	8.37		$7.53L, $8.37M per 2 beds, each use, w/ JD 2240		
Tractor cultivating 6x	7.56	3.48		1A at a time: 1 hour/A = 6 mins/2 beds; $1.26L, $0.53 +.05 = $0.58M per pass w/ Cub mostly		
Side-dressing				Spin 500 lbs 4-3-3/A, 1 hr total/20 beds = 6 mins/2 beds; $1.26L, $0.32 +.05 = $0.37M, $10Pr w/ Ford 4000		
Spraying				1 hr/.5A total time = 12 mins/2 beds; $2.51L, $0.64 +.10 = $0.74M, $6Pr w/ Ford 4000		
Flame weeding				10 beds/hr = 12 mins/2 beds; $2.51L, $0.64 +.10 = $0.74M, $6Pr w/ Ford 4000		
Other						
Pre-harvest Subtotal:	218.00	17.20	307.00	= 542.20 Pre-harvest cost for two beds		

Harvest:

Total yield for two 350' beds =	350 cases	24-count case, 2 bunches/plant
Total hours to harvest two 350' beds	140 hrs	at 60 bunches/hr

	$ Labor cost	$ Machinery cost	$ Product cost	Notes	
Field to pack house	1757.00			at $12.55/hr	140 hrs
Pack house to cooler	351.40			at $12.55/hr	at 300 bunches/hr: 28 hrs
Bags, boxes, labels			374.50	$0.25/bag, $1.00/box, $0.07/label	350 at $1.07
Delivery	30.12	9.60		See Worksheet 1.	
Post Harvest:					
Mow crop				6 beds at a time: 10 mins/2 beds; $2.09L, $0.53 +.17 = $0.70M w/ Ford 4000	
Remove mulch	12.55			1 hour/2 beds: $12.55L	
Disk	1.26	0.73		$1.26L, $0.63 +.10 = $0.73M w/ JD 2240, see disking above.	
Sow cover crop: spinner	1.26	0.68	8.00	1A at a time: 1 hr/20 beds = 6 mins/2 beds; $1.26L, $0.63 + .05 = $0.68M, $8Pr w/ JD 2240	
Sow cover crop: Brillion				1A at a time: 2 hrs/20 beds = 12 mins/2 beds; $2.51L, $1.26 + .20 = $1.46M, 8Pr w/ JD 2240	
Other					
Post-harvest Subtotal:	2371.59	28.21	689.50	= 3089.30 Harvested cost for 2 beds	

	$ Labor cost	$ Machinery cost	$ Product cost	Notes
Marketing Costs:				
Labor: sales calls for season (for this crop only)	6.28			Average 10 mins/week for 3 weeks: .5 hr
Commissions				Commissions, if any, to growers' co-op, broker, or salesperson
Farmers' market expense	60.24	4.70	9.00	See Worksheet 1.
Total Crop Costs:	2438.11	32.91	698.50	= 3169.52 Total crop costs
Overhead Costs:	288.00			Apportionment for two 350' beds, see Worksheet 1.
Total Costs: Crop & Overhead Total:	3457.52			Total costs per two 350' beds

Sales:	# of units	Price per unit	Total $
Retail:	75.00	36.00	2700.00
Wholesale:	275.00	20.00	5500.00
Other:			0.00
Total units	350.00		
Total Sales:			8200.00 For two 350' beds

Net Profit: Total sales – total costs =	4742.48	**Net profit for two 350' beds (1/10 acre)**
Net Profit/Acre:	47424.80	Standardize to one acre
Cost/Unit:	9.88	Total cost/total units
Net Profit/Unit:	13.55	Net profit/total units

NOTES:

Crop Enterprise Budget: Parsnips

Crop Enterprise Budget

Crop Year:		Crop: (and specify: early, mid, late)	Parsnips	Unit Area:	Two 350' beds
				Bed feet or acres:	700' or 1/10A
Today's Date:		Rows per bed & plant spacing:	3 rows/bed, 20 seeds/foot		
Costs in $:	Remember to prorate to unit area			Field:	

Note: Twenty 350' beds = 1 acre

NOTES: Labor at $12.55/hr. See Worksheet 1. Figures below are for two 350' beds.

	Labor cost $	Machinery cost $	Product cost $	Notes
Prepare Soil:				
Disk 1x	1.26	0.73		1A at a time: 1 hr total for 20 beds = 6 mins/2 beds; $1.26L, $0.63 + .10 = $0.73M w/ JD 2240; see Worksheet 4
Chisel 1x	2.51	0.74		.5A at a time:1 hr total for 10 beds = 12 mins/2 beds; $2.51L, $0.64 +.10 = $0.74M w/ Ford 4000; see Worksheet 4
Rototill 1x, 2x				.5A at a time: 2 hrs total for 10 beds = 24 mins/2 beds; $5.02L, $1.28 tractor + .52 tiller = $1.80M w/ Ford 4000
Bedform 2x	5.02	1.48		.5A at a time: 1 hr total for 10 beds = 12 mins/2 beds; $2.51L, $0.64 +.10 = $0.74M for ONE pass w/ Ford 4000
Fertilizer	1.26	0.68	10.00	500 lbs 4-3-3/A at a time: 1 hour total for 20 beds = 6 mins/2 beds; $.1.26L, $0.63 +.05 = $0.68M, $10Pr w/ JD 2240
Manure, compost	2.52	1.02	25.00	1A at a time: compost at $25/yd, 10 yds/A; 2 hrs total for 20 beds = 12 mins per 2 beds; $2.51L, $1.26 + .75 = $2.01M, $25Pr w/ JD 2240
Other				
Plastic mulch				.5A at a time: 1.5 hr/A laying = 10 mins/2 beds; $2.09L, $0.53 +.17 = $0.70M, $20Pr w/ Ford 4000
Seed/Transplant:				
Seeding in field	6.28		16.00	2 beds at a time: 30 mins/2 beds total = $6.28L 1/8 lb seed
Cost of transplants				$6.49/128 = $0.06/plant
Transplanting labor				3 rows by hand: 3 hrs/2 beds total = $37.65L 2 rows w/ transplanter, 6 beds at a time; 1 hr prep plants, 1.5hr x 3 people transplanting, 2 hrs machinery for 2 beds = $22.78L, $2.11 + .66 = $2.77M
Cultivation:				
Reemay on/off				For 2 beds: $105/3 uses = $35Pr, .75 hr laying = $9.41L
Hoeing 1x, 2x, 3x				at $12.55/hr: average 1 hr/2 beds $12.55/2 beds
Hand weeding 1	50.20			at $12.55/hr: average 8 hrs/2 beds $100.40/2 beds reduced by flaming
Hand weeding 2	25.10			at $12.55/hr: average 4 hrs/2 beds $50.20/2 beds
Hand weeding 3				at $12.55/hr: average 2 hrs/2 beds $25.10/2 beds
Straw mulch				40 bales at $3, 1 hr/2 beds; $12.55L, $120.00Pr
Irrigating 1x	7.53	8.37		$7.53L, $8.37M per 2 beds, each use, w/ JD 2240
Tractor cultivating 6x	7.56	3.48		1A at a time: 1 hour/A = 6 mins/2 beds; $1.26L, $0.53 +.05 = $0.58M per pass w/ Cub mostly
Side-dressing				Spin 500 lbs 4-3-3/A, 1 hr total/20 beds = 6 mins/2 beds; $1.26L, $0.32 +.05 = $0.37M, $10Pr w/ Ford 4000
Spraying				1 hr/.5A total time = 12 mins/2 beds; $2.51L, $0.64 +.10 = $0.74M, $6Pr w/ Ford 4000
Flame weeding	2.51	0.74	6.00	10 beds/hr = 12 mins/2 beds; $2.51L, $0.64 +.10 = $0.74M, $6Pr w/ Ford 4000
Other				
Pre-harvest Subtotal:	111.75	17.24	57.00	= 185.99 Pre-harvest cost for two beds

Harvest:

Total yield for two 350' beds = 60 bags

Total hours to harvest two 350' beds 12 hrs 5 bags/hr

	Labor cost $	Machinery cost $	Product cost $	Notes
Field to pack house	150.60			at $12.55/hr 12 hrs
Pack house to cooler	94.13			at $12.55/hr 8 bags/hr = 7.5 hrs
Bags, boxes, labels			15.00	$0.25/bag, $1.00/box, $0.07/label
Delivery	30.12	9.60		See Worksheet 1.
Post Harvest:				
Mow crop				6 beds at a time: 10 mins/2 beds; $2.09L, $0.53 +.17 = $0.70M w/ Ford 4000
Remove mulch				1 hour/2 beds: $12.55L
Disk	1.26	0.73		$1.26L, $0.63 +.10 = $0.73M w/ JD 2240, see disking above.
Sow cover crop: spinner	1.26	0.68	8.00	1A at a time: 1 hr/20 beds = 6 mins/2 beds; $1.26L, $0.63 + .05 = $0.68M, $8Pr w/ JD 2240
Sow cover crop: Brillion				1A at a time: 2 hrs/20 beds = 12 mins/2 beds; $2.51L, $1.26 + .20 = $1.46M, 8Pr w/ JD 2240
Other				
Post-harvest Subtotal:	389.12	28.25	80.00	= 497.37 Harvested cost for 2 beds

NOTES: Flaming reduces overall weedi

	Labor cost $	Machinery cost $	Product cost $	Notes
Marketing Costs:				
Labor: sales calls for season (for this crop only)	6.28			Average 10 mins/week for 3 weeks: .5 hr
Commissions				Commissions, if any, to growers' co-op, broker, or salesperson
Farmers' market expense	60.24	4.70	9.00	See Worksheet 1.
Total Crop Costs:	455.64	32.95	89.00	= 577.59 Total crop costs
Overhead Costs:	288.00			Apportionment for two 350' beds, see Worksheet 1.
Total Costs: Crop & Overhead Total:	865.59			Total costs per two 350' beds

Sales:	# of units	Price per unit	Total $
Retail:	10.00	50.00	500.00
Wholesale:	50.00	35.00	1750.00
Other:			0.00
Total units	60.00		
Total Sales:			2250.00 For two 350' beds

Net Profit: Total sales – total costs =	1384.41	**Net profit for two 350' beds (1/10 acre)**
Net Profit/Acre:	13844.10	Standardize to one acre
Cost/Unit:	14.43	Total cost/total units
Net Profit/Unit:	23.07	Net profit/total units

Crop Enterprise Budget: Peas (snap)

Crop Enterprise Budget

Crop Year: | Crop: **Peas: snap** (and specify: early, mid, late) | Unit Area: **Two 350' beds** | Note: Twenty 350' beds = 1 acre

Bed feet or acres: **700' or 1/10A**

Today's Date: | Rows per bed & plant spacing: 2 rows/bed

Costs in $: Remember to prorate to unit area | Field:

NOTES: Labor at $12.55/hr. See Worksheet 1. Figures below are for two 350' beds

	$ Labor cost	$ Machinery cost	$ Product cost	Notes
Prepare Soil:				
Disk 1x	1.26	0.73		1A at a time: 1 hr total for 20 beds = 6 mins/2 beds; $1.26L, $0.63 + .10 = $0.73M w/ JD 2240; see ' w/ JD, See Worksheet 4
Chisel 1x	2.51	0.74		.5A at a time:1 hr total for 10 beds = 12 mins/2 beds; $2.51L, $0.64 +.10 = $0.74M w/ Ford 4000; se w/ Ford 4000
Rototill 1x, 2x				.5A at a time: 2 hrs total for 10 beds = 24 mins/2 beds; $5.02L, $1.28 tractor + .52 tiller = $1.80M w/ Ford 4000 w/ Ford 4000
Bedform 2x	5.02	1.48		.5A at a time: 1 hr total for 10 beds = 12 mins/2 beds; $2.51L, $0.64 +.10 = $0.74M for ONE pass w/ Ford 4000 w/ Ford 4000
Fertilizer	1.26	0.68	10.00	500 lbs 4-3-3/A at a time: 1 hour total for 20 beds = 6 mins/2 beds; $.1.26L, $0.63 +.05 = $0.68M, $10Pr w/ JD 2240
Manure, compost	2.52	1.02	25.00	1A at a time: compost at $25/yd, 10 yds/A; 2 hrs total for 20 beds = 12 mins per 2 beds; $2.51L, $1.26 + .75 = $2.01M, $25Pr w/ JD 2240
Other				
Plastic mulch				.5A at a time: 1.5 hr/A laying = 10 mins/2 beds; $2.09L, $0.53 +.17 = $0.70M, $20Pr w/ Ford 4000
Seed/Transplant:				
Seeding in field	6.28		135.00	2 beds at a time: 30 mins/2 beds total = $6.28L 15 lbs seed
Cost of transplants				$6.49/128 = $0.06/plant
Transplanting labor				3 rows by hand: 3 hrs/2 beds total = $37.65L
				2 rows w/ transplanter, 6 beds at a time; 1 hr prep plants, 1.5hr x 3 people transplanting, 2 hrs machinery for 2 beds = $22.78L, $2.11 + .66 = $2.77M
Cultivation:				
Reemay on/off	9.41		35.00	For 2 beds: $105/3 uses = $35Pr, .75 hr laying = $9.41L
Hoeing 1x, 2x, 3x				at $12.55/hr: average 1 hr/2 beds $12.55/2 beds
Hand weeding 1	50.20			at $12.55/hr: average 8 hrs/2 beds $100.40/2 beds
Hand weeding 2	25.10			at $12.55/hr: average 4 hrs/2 beds $50.20/2 beds
Hand weeding 3				at $12.55/hr: average 2 hrs/2 beds $25.10/2 beds
Straw mulch				40 bales at $3, 1 hr/2 beds; $12.55L, $120.00Pr
Irrigating 1x	7.53	8.37		$7.53L, $8.37M per 2 beds, each use, w/ JD 2240
Tractor cultivating 6x	7.56	3.48		1A at a time: 1 hour/A = 6 mins/2 beds; $1.26L, $0.53 +.05 = $0.58M per pass w/ Cub mostly
Side-dressing				Spin 500 lbs 4-3-3/A, 1 hr total/20 beds = 6 mins/2 beds; $1.26L, $0.32 +.05 = $0.37M, $10Pr w/ Ford 4000
Spraying				1 hr/.5A total time = 12 mins/2 beds; $2.51L, $0.64 +.10 = $0.74M, $6Pr w/ Ford 4000
Flame weeding				10 beds/hr = 12 mins/2 beds; $2.51L, $0.64 +.10 = $0.74M, $6Pr w/ Ford 4000
Other				
Pre-harvest Subtotal:	118.65	16.50	205.00	= 340.15 Pre-harvest cost for two beds

Harvest:

Total yield for two 350' beds = 320 lbs

Total hours to harvest two 350' beds 12 hrs — 3 pickings

	Labor cost	Machinery cost	Product cost	Notes
Field to pack house	150.60			at $12.55/hr 12 hrs
Pack house to cooler	25.10			at $12.55/hr 2 hrs
Bags, boxes, labels			8.00	$0.25/bag, $1.00/box, $0.07/label 32 10-lb bags at $0.25
Delivery	30.12	9.60		See Worksheet 1.
Post Harvest:				
Mow crop	2.09	0.70		6 beds at a time: 10 mins/2 beds; $2.09L, $0.53 +.17 = $0.70M w/ Ford 4000
Remove mulch				1 hour/2 beds: $12.55L
Disk	1.26	0.73		$1.26L, $0.63 +.10 = $0.73M w/ JD 2240, see disking above.
Sow cover crop: spinner	1.26	0.68	8.00	1A at a time: 1 hr/20 beds = 6 mins/2 beds; $1.26L, $0.63 + .05 = $0.68M, $8Pr w/ JD 2240
Sow cover crop: Brillion				1A at a time: 2 hrs/20 beds = 12 mins/2 beds; $2.51L, $1.26 + .20 = $1.46M, 8Pr w/ JD 2240
Other				
Post-harvest Subtotal:	329.08	28.21	221.00	= 578.29 Harvested cost for 2 beds
Marketing Costs:				
Labor: sales calls for season (for this crop only)	6.28			Average 10 mins/week for 3 weeks: .5 hr
Commissions				Commissions, if any, to growers' co-op, broker, or salesperson
Farmers' market expense	60.24	4.70	9.00	See Worksheet 1.
Total Crop Costs:	395.60	32.91	230.00	= 658.51 Total crop costs
Overhead Costs:	288.00			Apportionment for two 350' beds, see Worksheet 1.
Total Costs: Crop & Overhead Total:	946.51			Total costs per two 350' beds

Sales:	# of units	Price per unit	Total $
Retail:	200.00	2.75	550.00
Wholesale:	120.00	1.50	180.00
Other:			0.00
Total units	320.00		
Total Sales:			730.00 For two 350' beds

Net Profit:		
Total sales – total costs =	-216.51	**Net profit for two 350' beds (1/10 acre)**
Net Profit/Acre:	-2165.10	Standardize to one acre
Cost/Unit:	2.96	Total cost/total units
Net Profit/Unit:	-0.68	Net profit/total units

NOTES:

Crop Enterprise Budget: Peppers (bell)

Crop Enterprise Budget

Crop Year:		Crop: (and specify: early, mid, late)	Peppers: bell
Unit Area:	Two 350' beds	Note: Twenty 350' beds = 1 acre	
Bed feet or acres:	700' or 1/10A		
Today's Date:		Rows per bed & plant spacing:	2 rows/bed, 16' apart, on plastic mulch
Field:			

Costs in $: Remember to prorate to unit area

NOTES: Labor at $12.55/hr. See Worksheet 1. Figures below are for two 350' beds

	Labor cost $	Machinery cost $	Product cost $	Notes
Prepare Soil:				
Disk 1x	1.26	0.73		1A at a time: 1 hr total for 20 beds = 6 mins/2 beds; $1.26L, $0.63 + .10 = $0.73M w/ JD 2240; see Workst w/ JD, See Worksheet 4
Chisel 1x	2.51	0.74		.5A at a time:1 hr total for 10 beds = 12 mins/2 beds; $2.51L, $0.64 +.10 = $0.74M w/ Ford 4000; see Worl w/ Ford 4000
Rototill 1x, 2x				.5A at a time: 2 hrs total for 10 beds = 24 mins/2 beds; $5.02L, $1.28 tractor + .52 tiller = $1.80M w/ Ford 4000 w/ Ford 4000
Bedform 2x	5.02	1.48		.5A at a time: 1 hr total for 10 beds = 12 mins/2 beds; $2.51L, $0.64 +.10 = $0.74M for ONE pass w/ Ford 4000 w/ Ford 4000
Fertilizer	1.26	0.68	10.00	500 lbs 4-3-3/A at a time: 1 hour total for 20 beds = 6 mins/2 beds; $.1.26L, $0.63 +.05 = $0.68M, $10Pr w/ JD 2240
Manure, compost	2.52	1.02	25.00	1A at a time: compost at $25/yd, 10 yds/A; 2 hrs total for 20 beds = 12 mins per 2 beds; $2.51L, $1.26 + .75 = $2.01M, $25Pr w/ JD 2240
Other				
Plastic mulch	2.09	0.70	20.00	.5A at a time: 1.5 hr/A laying = 10 mins/2 beds; $2.09L, $0.53 +.17 = $0.70M, $20Pr w/ Ford 4000
Seed/Transplant:				
Seeding in field				2 beds at a time: 30 mins/2 beds total = $6.28L
Cost of transplants			210.00	$6.49/128 = $0.06/plant; 1050 plants in 804s: $0.20/plant
Transplanting labor	37.65			3 rows by hand: 3 hrs/2 beds total = $37.65L; 2 rows but slower through plastic
				2 rows w/ transplanter, 6 beds at a time; 1 hr prep plants, 1.5hr x 3 people transplanting, 2 hrs machinery for 2 beds = $22.78L, $2.11 + .66 = $2.77M
Cultivation:				
Reemay on/off	9.41		35.00	For 2 beds: $105/3 uses = $35Pr, .75 hr laying = $9.41L
Hoeing 1x, 2x, 3x	25.10			at $12.55/hr: average 1 hr/2 beds; $12.55/2 beds
Hand weeding 1	12.55			at $12.55/hr: average 8 hrs/2 beds; $100.40/2 beds
Hand weeding 2	12.55			at $12.55/hr: average 4 hrs/2 beds; $50.20/2 beds
Hand weeding 3				at $12.55/hr: average 2 hrs/2 beds; $25.10/2 beds
Straw mulch				40 bales at $3, 1 hr/2 beds; $12.55L, $120.00Pr
Irrigating 1x	7.53	8.37		$7.53L, $8.37M per 2 beds, each use, w/ JD 2240
Tractor cultivating 6x	7.56	3.48		1A at a time: 1 hour/A = 6 mins/2 beds; $1.26L, $0.53 +.05 = $0.58M per pass w/ Cub mostly
Side-dressing				Spin 500 lbs 4-3-3/A, 1 hr total/20 beds = 6 mins/2 beds; $1.26L, $0.32 +.05 = $0.37M, $10Pr w/ Ford 4000
Spraying				1 hr/.5A total time = 12 mins/2 beds; $2.51L, $0.64 +.10 = $0.74M, $6Pr w/ Ford 4000
Flame weeding				10 beds/hr = 12 mins/2 beds; $2.51L, $0.64 +.10 = $0.74M, $6Pr w/ Ford 4000
Other				
Pre-harvest Subtotal:	127.01	17.20	300.00	= 444.21 Pre-harvest cost for two beds

Harvest:

Total yield for two 350' beds = 60 22-lb cases

Total hours to harvest two 350' beds 12 hrs — 12 pickings at 1 hr each

	Labor cost $	Machinery cost $	Product cost $	Notes
Field to pack house	150.60			at $12.55/hr; 12 hrs
Pack house to cooler	50.20			at $12.55/hr; at 15 cases/hr = 4 hrs
Bags, boxes, labels			64.20	$0.25/bag, $1.00/box, $0.07/label; 60 at $1.07
Delivery	30.12	9.60		See Worksheet 1.
Post Harvest:				
Mow crop	2.09	0.70		6 beds at a time: 10 mins/2 beds; $2.09L, $0.53 +.17 = $0.70M w/ Ford 4000
Remove mulch	12.55			1 hour/2 beds: $12.55L
Disk	1.26	0.73		$1.26L, $0.63 +.10 = $0.73M w/ JD 2240, see disking above.
Sow cover crop: spinner	1.26	0.68	8.00	1A at a time: 1 hr/20 beds = 6 mins/2 beds; $1.26L, $0.63 + .05 = $0.68M, $8Pr w/ JD 2240
Sow cover crop: Brillion				1A at a time: 2 hrs/20 beds = 12 mins/2 beds; $2.51L, $1.26 + .20 = $1.46M, 8Pr w/ JD 2240
Other				
Post-harvest Subtotal:	375.09	28.91	372.20	= 776.20 Harvested cost for 2 beds
Marketing Costs:				
Labor: sales calls for season (for this crop only)	6.28			Average 10 mins/week for 3 weeks: .5 hr
Commissions				Commissions, if any, to growers' co-op, broker, or salesperson
Farmers' market expense	60.24	4.70	9.00	See Worksheet 1.
Total Crop Costs:	441.61	33.61	381.20	= 856.42 Total crop costs
Overhead Costs:	288.00			Apportionment for two 350' beds, see Worksheet 1.
Total Costs: Crop & Overhead Total:	1144.42			Total costs per two 350' beds

Sales:

	# of units	Price per unit	Total $
Retail:	30.00	60.00	1800.00
Wholesale:	30.00	30.00	900.00
Other:			0.00
Total units	60.00		
Total Sales:			2700.00 For two 350' beds

Net Profit: Total sales – total costs =	1555.58	**Net profit for two 350' beds (1/10 acre)**
Net Profit/Acre:	15555.80	Standardize to one acre
Cost/Unit:	19.07	Total cost/total units
Net Profit/Unit:	25.93	Net profit/total units

NOTES:

Crop Enterprise Budget: Potatoes

Crop Enterprise Budget

Crop Year:		Crop: (and specify: early, mid, late)	Potatoes
Today's Date:		Rows per bed & plant spacing:	1 row/bed
Unit Area:	Two 350' beds	Note: Twenty 350' beds = 1 acre	
Bed feet or acres:	700' or 1/10A		
Field:			

Costs in $: Remember to prorate to unit area

NOTES: Labor at $12.55/hr. See Worksheet 1. Figures below are for two 350' beds.

	Labor cost $	Machinery cost $	Product cost $	Notes
Prepare Soil:				
Disk 1x	1.26	0.73		1A at a time: 1 hr total for 20 beds = 6 mins/2 beds; $1.26L, $0.63 + .10 = $0.73M w/ JD 2240; see V w/ JD, See Worksheet 4
Chisel 1x	2.51	0.74		.5A at a time:1 hr total for 10 beds = 12 mins/2 beds; $2.51L, $0.64 +.10 = $0.74M w/ Ford 4000; see w/ Ford 4000
Rototill 1x, 2x				.5A at a time: 2 hrs total for 10 beds = 24 mins/2 beds; $5.02L, $1.28 tractor + .52 tiller = $1.80M w/ Ford 4000 w/ Ford 4000
Bedform 2x	2.51	0.74		.5A at a time: 1 hr total for 10 beds = 12 mins/2 beds; $2.51L, $0.64 +.10 = $0.74M for ONE pass w/ Ford 4000 w/ Ford 4000
Fertilizer	1.26	0.68	10.00	500 lbs 4-3-3/A at a time: 1 hour total for 20 beds = 6 mins/2 beds; $.1.26L, $0.63 +.05 = $0.68M, $10Pr w/ JD 2240
Manure, compost	2.52	1.02	25.00	1A at a time: compost at $25/yd, 10 yds/A; 2 hrs total for 20 beds = 12 mins per 2 beds; $2.51L, $1.26 + .75 = $2.01M, $25Pr w/ JD 2240
Other	2.51	0.74		
Plastic mulch				.5A at a time: 1.5 hr/A laying = 10 mins/2 beds; $2.09L, $0.53 +.17 = $0.70M, $20Pr w/ Ford 4000
Seed/Transplant:				
Seeding in field	25.10		160.00	100 lbs seed, 2 hrs cut and drop in furrow
Cost of transplants				$6.49/128 = $0.06/plant
Transplanting labor				3 rows by hand: 3 hrs/2 beds total = $37.65L
				2 rows w/ transplanter, 6 beds at a time; 1 hr prep plants, 1.5hr x 3 people transplanting, 2 hrs machinery for 2 beds = $22.78L, $2.11 + .66 = $2.77M
Cultivation:				
Reemay on/off				For 2 beds: $105/3 uses = $35Pr, .75 hr laying = $9.41L
Hoeing 1x, 2x, 3x				at $12.55/hr: average 1 hr/2 beds $12.55/2 beds
Hand weeding 1	12.55			at $12.55/hr: average 8 hrs/2 beds $100.40/2 beds
Hand weeding 2	12.55			at $12.55/hr: average 4 hrs/2 beds $50.20/2 beds
Hand weeding 3	12.55			at $12.55/hr: average 2 hrs/2 beds $25.10/2 beds
Straw mulch				40 bales at $3, 1 hr/2 beds; $12.55L, $120.00Pr
Irrigating 1x	7.53	8.37		$7.53L, $8.37M per 2 beds, each use, w/ JD 2240
Tractor cultivating 6x	7.56	3.48		1A at a time: 1 hour/A = 6 mins/2 beds; $1.26L, $0.53 +.05 = $0.58M per pass w/ Cub mostly
Side-dressing				Spin 500 lbs 4-3-3/A, 1 hr total/20 beds = 6 mins/2 beds; $1.26L, $0.32 +.05 = $0.37M, $10Pr w/ Ford 4000
Spraying	5.02	1.48	36.00	1 hr/.5A total time = 12 mins/2 beds; $2.51L, $0.64 +.10 = $0.74M, $6Pr w/ Ford 4000
Flame weeding	2.51	0.74	6.00	10 beds/hr = 12 mins/2 beds; $2.51L, $0.64 +.10 = $0.74M, $6Pr w/ Ford 4000
Other	2.09	0.70		
Pre-harvest Subtotal:	100.03	19.42	237.00	= 356.45 Pre-harvest cost for two beds

Harvest:

Total yield for two 350' beds = 25 50-lb cases

Total hours to harvest two 350' beds 5 hrs 5 cases/hr

	Labor cost $	Machinery cost $	Product cost $	Notes
Field to pack house	69.03	4.15		at $12.55/hr 5 hrs picking plus .5 hr chain digging: 4.15M
Pack house to cooler	62.75			at $12.55/hr at 5 cases/hr: 5 hrs
Bags, boxes, labels			26.75	$0.25/bag, $1.00/box, $0.07/label 25 at $1.07
Delivery	30.12	9.60		See Worksheet 1.
Post Harvest:				
Mow crop				6 beds at a time: 10 mins/2 beds; $2.09L, $0.53 +.17 = $0.70M w/ Ford 4000
Remove mulch				1 hour/2 beds: $12.55L
Disk	1.26	0.73		$1.26L, $0.63 +.10 = $0.73M w/ JD 2240, see disking above.
Sow cover crop: spinner	1.26	0.68	8.00	1A at a time: 1 hr/20 beds = 6 mins/2 beds; $1.26L, $0.63 + .05 = $0.68M, $8Pr w/ JD 2240
Sow cover crop: Brillion				1A at a time: 2 hrs/20 beds = 12 mins/2 beds; $2.51L, $1.26 + .20 = $1.46M, 8Pr w/ JD 2240
Other				
Post-harvest Subtotal:	264.45	34.58	271.75	= 570.78 Harvested cost for 2 beds
Marketing Costs:				
Labor: sales calls for season (for this crop only)	6.28			Average 10 mins/week for 3 weeks: .5 hr
Commissions				Commissions, if any, to growers' co-op, broker, or salesperson
Farmers' market expense	60.24	4.70	9.00	See Worksheet 1.
Total Crop Costs:	330.97	39.28	280.75	= 651.00 Total crop costs
Overhead Costs:	288.00			Apportionment for two 350' beds, see Worksheet 1.
Total Costs: Crop & Overhead Total:	939.00			Total costs per two 350' beds

Sales:	# of units	Price per unit	Total $
Retail:	10.00	75.00	750.00
Wholesale:	15.00	30.00	450.00
Other:			0.00
Total units	25.00		
Total Sales:			1200.00 For two 350' beds

Net Profit: Total sales – total costs =	261.00	**Net profit for two 350' beds (1/10 acre)**
Net Profit/Acre:	2610.00	Standardize to one acre
Cost/Unit:	37.56	Total cost/total units
Net Profit/Unit:	10.44	Net profit/total units

NOTES:

Crop Enterprise Budget: Spinach

Crop Enterprise Budget

Crop Year: | Crop: **Spinach** (and specify: early, mid, late) | **Unit Area:** **Two 350' beds** | **Note: Twenty 350' beds = 1 acre**

Bed feet or acres: **700' or 1/10A**

Today's Date: | Rows per bed & plant spacing: 3 rows/bed, direct seeded

Costs in $: Remember to prorate to unit area | Field:

NOTES: Labor at $12.55/hr. See Worksheet 1. Figures below are for two 350' beds.

	Labor cost $	Machinery cost $	Product cost $	Notes
Prepare Soil:				
Disk 1x	1.26	0.73		1A at a time: 1 hr total for 20 beds = 6 mins/2 beds; $1.26L, $0.63 + .10 = $0.73M w/ JD 2240; see Worksheet 4
Chisel 1x	2.51	0.74		.5A at a time:1 hr total for 10 beds = 12 mins/2 beds; $2.51L, $0.64 +.10 = $0.74M w/ Ford 4000; see Worksheet 4
Rototill 1x, 2x				.5A at a time: 2 hrs total for 10 beds = 24 mins/2 beds; $5.02L, $1.28 tractor + .52 tiller = $1.80M w/ Ford 4000
Bedform 2x	5.02	1.48		.5A at a time: 1 hr total for 10 beds = 12 mins/2 beds; $2.51L, $0.64 +.10 = $0.74M for ONE pass w/ Ford 4000
Fertilizer	1.26	0.68	10.00	500 lbs 4-3-3/A at a time: 1 hour total for 20 beds = 6 mins/2 beds; $.1.26L, $0.63 +.05 = $0.68M, $10Pr w/ JD 2240
Manure, compost	2.52	1.02	25.00	1A at a time: compost at $25/yd, 10 yds/A; 2 hrs total for 20 beds = 12 mins per 2 beds; $2.51L, $1.26 + .75 = $2.01M, $25Pr w/ JD 2240
Other				
Plastic mulch				.5A at a time: 1.5 hr/A laying = 10 mins/2 beds; $2.09L, $0.53 +.17 = $0.70M, $20Pr w/ Ford 4000
Seed/Transplant:				
Seeding in field	6.28		14.70	2 beds at a time: 30 mins/2 beds total = $6.28L 21,000 seeds
Cost of transplants				$6.49/128 = $0.06/plant
Transplanting labor				3 rows by hand: 3 hrs/2 beds total = $37.65L
				2 rows w/ transplanter, 6 beds at a time; 1 hr prep plants, 1.5hr x 3 people transplanting, 2 hrs machinery for 2 beds = $22.78L, $2.11 + .66 = $2.77M
Cultivation:				
Reemay on/off				For 2 beds: $105/3 uses = $35Pr, .75 hr laying = $9.41L
Hoeing 1x, 2x, 3x				at $12.55/hr: average 1 hr/2 beds $12.55/2 beds
Hand weeding 1	50.20			at $12.55/hr: average 8 hrs/2 beds $100.40/2 beds easier to weed
Hand weeding 2				at $12.55/hr: average 4 hrs/2 beds $50.20/2 beds
Hand weeding 3				at $12.55/hr: average 2 hrs/2 beds $25.10/2 beds
Straw mulch				40 bales at $3, 1 hr/2 beds; $12.55L, $120.00Pr
Irrigating 1x	7.53	8.37		$7.53L, $8.37M per 2 beds, each use, w/ JD 2240
Tractor cultivating 6x	7.56	3.48		1A at a time: 1 hour/A = 6 mins/2 beds; $1.26L, $0.53 +.05 = $0.58M per pass w/ Cub mostly
Side-dressing				Spin 500 lbs 4-3-3/A, 1 hr total/20 beds = 6 mins/2 beds; $1.26L, $0.32 +.05 = $0.37M, $10Pr w/ Ford 4000
Spraying				1 hr/.5A total time = 12 mins/2 beds; $2.51L, $0.64 +.10 = $0.74M, $6Pr w/ Ford 4000
Flame weeding				10 beds/hr = 12 mins/2 beds; $2.51L, $0.64 +.10 = $0.74M, $6Pr w/ Ford 4000
Other				
Pre-harvest Subtotal:	84.14	16.50	49.70	= 150.34 Pre-harvest cost for two beds

Harvest:

Total yield for two 350' beds = 700 lbs — 1 lb per bedfoot

Total hours to harvest two 350' beds 23.3 hrs — average 30 lbs per hour

	Labor cost $	Machinery cost $	Product cost $	Notes
Field to pack house	292.42			at $12.55/hr 23.3 hrs
Pack house to cooler	72.79			at $12.55/hr washing: 120 lbs per hour = 5.8 hrs
Bags, boxes, labels			74.90	$0.25/bag, $1.00/box, $0.07/label 70 boxes at $1.07
Delivery	30.12	9.60		See Worksheet 1.
Post Harvest:				
Mow crop				6 beds at a time: 10 mins/2 beds; $2.09L, $0.53 +.17 = $0.70M w/ Ford 4000
Remove mulch				1 hour/2 beds: $12.55L
Disk	1.26	0.73		$1.26L, $0.63 +.10 = $0.73M w/ JD 2240, see disking above.
Sow cover crop: spinner	1.26	0.68	8.00	1A at a time: 1 hr/20 beds = 6 mins/2 beds; $1.26L, $0.63 + .05 = $0.68M, $8Pr w/ JD 2240
Sow cover crop: Brillion				1A at a time: 2 hrs/20 beds = 12 mins/2 beds; $2.51L, $1.26 + .20 = $1.46M, 8Pr w/ JD 2240
Other				
Post-harvest Subtotal:	481.99	27.51	132.60	= 642.10 Harvested cost for 2 beds
Marketing Costs:				
Labor: sales calls for season (for this crop only)	6.28			Average 10 mins/week for 3 weeks: .5 hr
Commissions				Commissions, if any, to growers' co-op, broker, or salesperson
Farmers' market expense	60.24	4.70	9.00	See Worksheet 1.
Total Crop Costs:	548.51	32.21	141.60	= 722.32 Total crop costs
Overhead Costs:	288.00			Apportionment for two 350' beds, see Worksheet 1.
Total Costs: Crop & Overhead Total:	1010.32			Total costs per two 350' beds

Sales:

	# of units	Price per unit	Total $
Retail:	200.00	4.50	900.00
Wholesale:	500.00	2.25	1125.00
Other:			0.00
Total units	700.00		
Total Sales:			2025.00 For two 350' beds

Net Profit: Total sales – total costs =	1014.68	**Net profit for two 350' beds (1/10 acre)**
Net Profit/Acre:	10146.80	Standardize to one acre
Cost/Unit:	1.44	Total cost/total units
Net Profit/Unit:	1.45	Net profit/total units

NOTES:

Crop Enterprise Budget: Squash (summer)

Crop Enterprise Budget

Crop Year: | Crop: **Squash: summer** (and specify: early, mid, late) | **Unit Area:** **Two 350' beds** | **Note: Twenty 350' beds = 1 acre**

Bed feet or acres: **700' or 1/10A**

Today's Date: | Rows per bed & plant spacing: 1 row/bed, 2' in row, on plastic mulch

Costs in $: Remember to prorate to unit area | Field:

	$ Labor cost	$ Machinery cost	$ Product cost	NOTES: Labor at $12.55/hr. See Worksheet 1. Figures below are for two 350' beds.
Prepare Soil:				
Disk 1x	1.26	0.73		1A at a time: 1 hr total for 20 beds = 6 mins/2 beds; $1.26L, $0.63 + .10 = $0.73M w/ JD 2240; see Worksheet 4
Chisel 1x	2.51	0.74		.5A at a time:1 hr total for 10 beds = 12 mins/2 beds; $2.51L, $0.64 +.10 = $0.74M w/ Ford 4000; see Worksheet 4
Rototill 1x, 2x				.5A at a time: 2 hrs total for 10 beds = 24 mins/2 beds; $5.02L, $1.28 tractor + .52 tiller = $1.80M w/ Ford 4000
Bedform 2x	5.02	1.48		.5A at a time: 1 hr total for 10 beds = 12 mins/2 beds; $2.51L, $0.64 +.10 = $0.74M for ONE pass w/ Ford 4000
Fertilizer	1.26	0.68	10.00	500 lbs 4-3-3/A at a time: 1 hour total for 20 beds = 6 mins/2 beds; $.1.26L, $0.63 +.05 = $0.68M, $10Pr w/ JD 2240
Manure, compost	2.52	1.02	25.00	1A at a time: compost at $25/yd, 10 yds/A; 2 hrs total for 20 beds = 12 mins per 2 beds; $2.51L, $1.26 + .75 = $2.01M, $25Pr w/ JD 2240
Other				
Plastic mulch	2.09	0.70	20.00	.5A at a time: 1.5 hr/A laying = 10 mins/2 beds; $2.09L, $0.53 +.17 = $0.70M, $20Pr w/ Ford 4000
Seed/Transplant:				
Seeding in field				2 beds at a time: 30 mins/2 beds total = $6.28L
Cost of transplants			70.00	$6.49/128 = $0.06/plant; 350 plants in 804s: $0.20/plant
Transplanting labor	25.10			3 rows by hand: 3 hrs/2 beds total = $37.65L; 2 hrs total; 2 rows w/ transplanter, 6 beds at a time; 1 hr prep plants, 1.5hr x 3 people transplanting, 2 hrs machinery for 2 beds = $22.78L, $2.11 + .66 = $2.77M
Cultivation:				
Reemay on/off	9.41		35.00	For 2 beds: $105/3 uses = $35Pr, .75 hr laying = $9.41L
Hoeing 1x, 2x, 3x	25.10			at $12.55/hr: average 1 hr/2 beds; $12.55/2 beds
Hand weeding 1	12.55			at $12.55/hr: average 8 hrs/2 beds; $100.40/2 beds
Hand weeding 2	12.55			at $12.55/hr: average 4 hrs/2 beds; $50.20/2 beds
Hand weeding 3				at $12.55/hr: average 2 hrs/2 beds; $25.10/2 beds
Straw mulch				40 bales at $3, 1 hr/2 beds; $12.55L, $120.00Pr
Irrigating 1x	7.53	8.37		$7.53L, $8.37M per 2 beds, each use, w/ JD 2240
Tractor cultivating 6x	7.56	3.48		1A at a time: 1 hour/A = 6 mins/2 beds; $1.26L, $0.53 +.05 = $0.58M per pass w/ Cub mostly
Side-dressing				Spin 500 lbs 4-3-3/A, 1 hr total/20 beds = 6 mins/2 beds; $1.26L, $0.32 +.05 = $0.37M, $10Pr w/ Ford 4000
Spraying				1 hr/.5A total time = 12 mins/2 beds; $2.51L, $0.64 +.10 = $0.74M, $6Pr w/ Ford 4000
Flame weeding				10 beds/hr = 12 mins/2 beds; $2.51L, $0.64 +.10 = $0.74M, $6Pr w/ Ford 4000
Other				
Pre-harvest Subtotal:	114.46	17.20	160.00	= 291.66 Pre-harvest cost for two beds

Harvest:

Total yield for two 350' beds = 60 20-lb cases

Total hours to harvest two 350' beds 15 hrs — 20 pickings at .75 hr average

	Labor cost	Machinery cost	Product cost	Notes
Field to pack house	188.25			at $12.55/hr; 15 hrs
Pack house to cooler	50.20			at $12.55/hr; at 15 cases/hr: 4 hrs
Bags, boxes, labels			48.00	$0.25/bag, $1.00/box, $0.07/label; 60 boxes at $0.80
Delivery	30.12	9.60		See Worksheet 1.
Post Harvest:				
Mow crop	2.09	0.70		6 beds at a time: 10 mins/2 beds; $2.09L, $0.53 +.17 = $0.70M w/ Ford 4000
Remove mulch	12.55			1 hour/2 beds: $12.55L
Disk	1.26	0.73		$1.26L, $0.63 +.10 = $0.73M w/ JD 2240, see disking above.
Sow cover crop: spinner	1.26	0.68	8.00	1A at a time: 1 hr/20 beds = 6 mins/2 beds; $1.26L, $0.63 + .05 = $0.68M, $8Pr w/ JD 2240
Sow cover crop: Brillion				1A at a time: 2 hrs/20 beds = 12 mins/2 beds; $2.51L, $1.26 + .20 = $1.46M, 8Pr w/ JD 2240
Other				
Post-harvest Subtotal:	400.19	28.91	216.00	= 645.10 Harvested cost for 2 beds
Marketing Costs:				
Labor: sales calls for season (for this crop only)	6.28			Average 10 mins/week for 3 weeks: .5 hr
Commissions				Commissions, if any, to growers' co-op, broker, or salesperson
Farmers' market expense	60.24	4.70	9.00	See Worksheet 1
Total Crop Costs:	466.71	33.61	225.00	= 725.32 Total crop costs
Overhead Costs:	288.00			Apportionment for two 350' beds, see Worksheet 1.
Total Costs: Crop & Overhead Total:	1013.32			Total costs per two 350' beds

Sales:	# of units	Price per unit	Total $	
Retail:	20.00	40.00	800.00	
Wholesale:	40.00	25.00	1000.00	
Other:			0.00	
Total units	60.00			
Total Sales:			1800.00	For two 350' beds

Net Profit: Total sales – total costs =	786.68	**Net profit for two 350' beds (1/10 acre)**
Net Profit/Acre:	7866.80	Standardize to one acre
Cost/Unit:	16.89	Total cost/total units
Net Profit/Unit:	13.11	Net profit/total units

NOTES:

Crop Enterprise Budget: Squash (winter)

Crop Enterprise Budget

Crop Year: | Crop: **Squash: winter** (and specify: early, mid, late) | **Unit Area: Two 350' beds** | **Note: Twenty 350' beds = 1 acre**

Bed feet or acres: **700' or 1/10A**

Today's Date: | Rows per bed & plant spacing: 1 row/bed, 2 plants every 3', transplanted on mulch

Costs in $: Remember to prorate to unit area | Field:

NOTES: Labor at $12.55/hr. See Worksheet 1. Figures below are for two 350' beds.

	Labor cost $	Machinery cost $	Product cost $	Notes
Prepare Soil:				
Disk 1x	1.26	0.73		1A at a time: 1 hr total for 20 beds = 6 mins/2 beds; $1.26L, $0.63 + .10 = $0.73M w/ JD 2240; see Worksheet 4
Chisel 1x	2.51	0.74		.5A at a time:1 hr total for 10 beds = 12 mins/2 beds; $2.51L, $0.64 +.10 = $0.74M w/ Ford 4000; see Worksheet 4
Rototill 1x, 2x				.5A at a time: 2 hrs total for 10 beds = 24 mins/2 beds; $5.02L, $1.28 tractor + .52 tiller = $1.80M w/ Ford 4000
Bedform 2x	5.02	1.48		.5A at a time: 1 hr total for 10 beds = 12 mins/2 beds; $2.51L, $0.64 +.10 = $0.74M for ONE pass w/ Ford 4000
Fertilizer	1.26	0.68	10.00	500 lbs 4-3-3/A at a time: 1 hour total for 20 beds = 6 mins/2 beds; $.1.26L, $0.63 +.05 = $0.68M, $10Pr w/ JD 2240
Manure, compost	2.52	1.02	25.00	1A at a time: compost at $25/yd, 10 yds/A; 2 hrs total for 20 beds = 12 mins per 2 beds; $2.51L, $1.26 + .75 = $2.01M, $25Pr w/ JD 2240
Other				
Plastic mulch	2.09	0.70	20.00	.5A at a time: 1.5 hr/A laying = 10 mins/2 beds; $2.09L, $0.53 +.17 = $0.70M, $20Pr w/ Ford 4000
Seed/Transplant:				
Seeding in field				2 beds at a time: 30 mins/2 beds total = $6.28L
Cost of transplants			47.00	$6.49/128 = $0.06/plant; 235 plants (doubles) in 804s: $0.20/cell
Transplanting labor	25.10			3 rows by hand: 3 hrs/2 beds total = $37.65L; 2 hrs total; 2 rows w/ transplanter, 6 beds at a time; 1 hr prep plants, 1.5hr x 3 people transplanting, 2 hrs machinery for 2 beds = $22.78L, $2.11 + .66 = $2.77M
Cultivation:				
Reemay on/off				For 2 beds: $105/3 uses = $35Pr, .75 hr laying = $9.41L
Hoeing 1x, 2x, 3x	25.10			at $12.55/hr: average 1 hr/2 beds; $12.55/2 beds
Hand weeding 1	12.55			at $12.55/hr: average 8 hrs/2 beds; $100.40/2 beds
Hand weeding 2				at $12.55/hr: average 4 hrs/2 beds; $50.20/2 beds
Hand weeding 3				at $12.55/hr: average 2 hrs/2 beds; $25.10/2 beds
Straw mulch				40 bales at $3, 1 hr/2 beds; $12.55L, $120.00Pr
Irrigating 1x	7.53	8.37		$7.53L, $8.37M per 2 beds, each use, w/ JD 2240
Tractor cultivating 6x	7.56	3.48		1A at a time: 1 hour/A = 6 mins/2 beds; $1.26L, $0.53 +.05 = $0.58M per pass w/ Cub mostly
Side-dressing				Spin 500 lbs 4-3-3/A, 1 hr total/20 beds = 6 mins/2 beds; $1.26L, $0.32 +.05 = $0.37M, $10Pr w/ Ford 4000
Spraying				1 hr/.5A total time = 12 mins/2 beds; $2.51L, $0.64 +.10 = $0.74M, $6Pr w/ Ford 4000
Flame weeding				10 beds/hr = 12 mins/2 beds; $2.51L, $0.64 +.10 = $0.74M, $6Pr w/ Ford 4000
Other				
Pre-harvest Subtotal:	92.50	17.20	102.00	= 211.70 Pre-harvest cost for two beds

Harvest:

Total yield for two 350' beds = 30 45-lb cases

Total hours to harvest two 350' beds 7.5 hrs at 4 45-lb cases/hr

	Labor cost	Machinery cost	Product cost	Notes
Field to pack house	94.13			at $12.55/hr; 7.5 hrs
Pack house to cooler	62.75			at $12.55/hr; at 6 cases/hr = 5 hrs
Bags, boxes, labels			32.10	$0.25/bag, $1.00/box, $0.07/label; 30 boxes at $1.07
Delivery	30.12	9.60		See Worksheet 1.
Post Harvest:				
Mow crop				6 beds at a time: 10 mins/2 beds; $2.09L, $0.53 +.17 = $0.70M w/ Ford 4000
Remove mulch	12.55			1 hour/2 beds: $12.55L
Disk	1.26	0.73		$1.26L, $0.63 +.10 = $0.73M w/ JD 2240, see disking above.
Sow cover crop: spinner	1.26	0.68	8.00	1A at a time: 1 hr/20 beds = 6 mins/2 beds; $1.26L, $0.63 + .05 = $0.68M, $8Pr w/ JD 2240
Sow cover crop: Brillion				1A at a time: 2 hrs/20 beds = 12 mins/2 beds; $2.51L, $1.26 + .20 = $1.46M, 8Pr w/ JD 2240
Other				
Post-harvest Subtotal:	294.57	28.21	142.10	= 464.88 Harvested cost for 2 beds

	Labor cost	Machinery cost	Product cost	Notes
Marketing Costs:				
Labor: sales calls for season (for this crop only)	6.28			Average 10 mins/week for 3 weeks: .5 hr
Commissions				Commissions, if any, to growers' co-op, broker, or salesperson
Farmers' market expense	60.24	4.70	9.00	See Worksheet 1.
Total Crop Costs:	361.09	32.91	151.10	= 545.10 Total crop costs
Overhead Costs:	288.00			Apportionment for two 350' beds, see Worksheet 1.
Total Costs: **Crop & Overhead Total:**	833.10			Total costs per two 350' beds

Sales:	# of units	Price per unit	Total $
Retail:	10.00	40.00	400.00
Wholesale:	20.00	26.00	520.00
Other:			0.00
Total units	30.00		
Total Sales:			920.00 For two 350' beds

Net Profit: Total sales – total costs =	86.90	**Net profit for two 350' beds (1/10 acre)**
Net Profit/Acre:	869.00	Standardize to one acre
Cost/Unit:	27.77	Total cost/total units
Net Profit/Unit:	2.90	Net profit/total units

NOTES:

Crop Enterprise Budget: Tomatoes (field)

Crop Enterprise Budget

Crop Year: | Crop: Tomatoes: field (and specify: early, mid, late) | Unit Area: Two 350' beds | Note: Twenty 350' beds = 1 acre

Bed feet or acres: 700' or 1/10A

Today's Date: | Rows per bed & plant spacing: 1 row/bed, 3' spacing, transplanted on mulch

Costs in $: Remember to prorate to unit area | Field:

NOTES: Labor at $12.55/hr. See Worksheet 1. Figures below are for two 350' beds.

Costs in $	Labor cost $	Machinery cost $	Product cost $	Notes
Prepare Soil:				
Disk 1x	1.26	0.73		1A at a time: 1 hr total for 20 beds = 6 mins/2 beds; $1.26L, $0.63 + .10 = $0.73M w/ JD 2240; see Worksheet 4
Chisel 1x	2.51	0.74		.5A at a time:1 hr total for 10 beds = 12 mins/2 beds; $2.51L, $0.64 +.10 = $0.74M w/ Ford 4000; see Worksheet 4
Rototill 1x, 2x				.5A at a time: 2 hrs total for 10 beds = 24 mins/2 beds; $5.02L, $1.28 tractor + .52 tiller = $1.80M w/ Ford 4000
Bedform 2x	5.02	1.48		.5A at a time: 1 hr total for 10 beds = 12 mins/2 beds; $2.51L, $0.64 +.10 = $0.74M for ONE pass w/ Ford 4000
Fertilizer	1.26	0.68	10.00	500 lbs 4-3-3/A at a time: 1 hour total for 20 beds = 6 mins/2 beds; $.1.26L, $0.63 +.05 = $0.68M, $10Pr w/ JD 2240
Manure, compost	2.52	1.02	25.00	1A at a time: compost at $25/yd, 10 yds/A; 2 hrs total for 20 beds = 12 mins per 2 beds; $2.51L, $1.26 + .75 = $2.01M, $25Pr w/ JD 2240
Other				
Plastic mulch	2.09	0.70	20.00	.5A at a time: 1.5 hr/A laying = 10 mins/2 beds; $2.09L, $0.53 +.17 = $0.70M, $20Pr w/ Ford 4000
Seed/Transplant:				
Seeding in field				2 beds at a time: 30 mins/2 beds total = $6.28L
Cost of transplants			47.00	$6.49/128 = $0.06/plant; 235 plants in 804s: $0.20/plant
Transplanting labor	25.10			3 rows by hand: 3 hrs/2 beds total = $37.65L; 2 hrs total; 2 rows w/ transplanter, 6 beds at a time; 1 hr prep plants, 1.5hr x 3 people transplanting, 2 hrs machinery for 2 beds = $22.78L, $2.11 + .66 = $2.77M
Cultivation:				
Reemay on/off	9.41		35.00	For 2 beds: $105/3 uses = $35Pr, .75 hr laying = $9.41L
Hoeing 1x, 2x, 3x	25.10			at $12.55/hr: average 1 hr/2 beds; $12.55/2 beds
Hand weeding 1	12.55			at $12.55/hr: average 8 hrs/2 beds; $100.40/2 beds
Hand weeding 2	12.55			at $12.55/hr: average 4 hrs/2 beds; $50.20/2 beds
Hand weeding 3				at $12.55/hr: average 2 hrs/2 beds; $25.10/2 beds
Straw mulch				40 bales at $3, 1 hr/2 beds; $12.55L, $120.00Pr
Irrigating 1x	7.53	8.37		$7.53L, $8.37M per 2 beds, each use, w/ JD 2240
Tractor cultivating 3x	3.78	1.74		1A at a time: 1 hour/A = 6 mins/2 beds; $1.26L, $0.53 +.05 = $0.58M per pass w/ Cub mostly
Side-dressing				Spin 500 lbs 4-3-3/A, 1 hr total/20 beds = 6 mins/2 beds; $1.26L, $0.32 +.05 = $0.37M, $10Pr w/ Ford 4000
Spraying 4x	10.04	2.96	24.00	1 hr/.5A total time = 12 mins/2 beds; $2.51L, $0.64 +.10 = $0.74M, $6Pr w/ Ford 4000
Flame weeding				10 beds/hr = 12 mins/2 beds; $2.51L, $0.64 +.10 = $0.74M, $6Pr w/ Ford 4000
Staking and tieing	100.40		235.00	8 hrs total, 470 stakes/4 uses: $0.50/use
Pre-harvest Subtotal:	221.12	18.42	396.00	= 635.54 Pre-harvest cost for two beds

Harvest:

Total yield for two 350' beds = 100 20-lb boxes

Total hours to harvest two 350' beds: 24 hrs — 12 pickings at 2 hrs each

Item	Labor cost $	Machinery cost $	Product cost $	Notes
Field to pack house	301.20			at $12.55/hr; 24 hrs
Pack house to cooler	313.75			at $12.55/hr; at 4 cases/hr: 25 hrs
Bags, boxes, labels			157.00	$0.25/bag, $1.00/box, $0.07/label; 100 boxes at $1.57
Delivery	30.12	9.60		See Worksheet 1.
Post Harvest:				
Remove stakes, plants	37.65			3 hrs
Remove mulch	12.55			1 hour/2 beds: $12.55L
Disk	1.26	0.73		$1.26L, $.63 +.10 =$0.73M w/ JD, see disking above.
Sow cover crop: spinner	1.26	0.68	8.00	1A at a time: 1 hr/20 beds = 6 mins/2 beds; $1.26L, $0.63 + .05 = $0.68M, $8Pr, w/ JD 2240
Sow cover crop: Brillion				1A at a time: 2 hrs/20 beds = 12 mins/2 beds; $2.51L, $1.26 +.20 = $1.46M, 8Pr, w/ JD 2240
Other				
Post-harvest Subtotal:	918.91	29.43	561.00	= 1509.34 Harvested cost for 2 beds
Marketing Costs:				
Labor: sales calls for season (for this crop only)	6.28			Average 10 mins/week for 3 weeks: .5 hr
Commissions				Commissions, if any, to growers' co-op, broker, or salesperson
Farmers' market expense	60.24	4.70	9.00	See Worksheet 1
Total Crop Costs:	985.43	34.13	570.00	= 1589.56 Total crop costs
Overhead Costs:	288.00			Apportionment for two 350' beds, see Worksheet 1.
Total Costs: Crop & Overhead Total:	1877.56			Total costs per two 350' beds

Sales:	# of units	Price per unit	Total $
Retail:	30.00	55.00	1650.00
Wholesale:	70.00	30.00	2100.00
Other:			0.00
Total units	100.00		
Total Sales:			3750.00 For two 350' beds

Net Profit: Total sales – total costs =	1872.44	**Net profit for two 350' beds (1/10 acre)**
Net Profit/Acre:	18724.40	Standardize to one acre
Cost/Unit:	18.78	Total cost/total units
Net Profit/Unit:	18.72	Net profit/total units

NOTES:

Crop Enterprise Budget: Tomatoes (greenhouse)

Crop Enterprise Budget

Crop Year: | Crop: **Tomatoes: greenhouse** (and specify: early, mid, late) | Unit Area: **21' x 96' greenhouse**

Bed feet or acres:

Today's Date: | Rows per bed & plant spacing: 5 rows total, 12" in row spacing, non-grafted plants

Costs in $: Remember to prorate to unit area | Field:

NOTES: Labor at $12.55/hr. See Worksheet 1.

	$ Labor cost	$ Machinery cost	$ Product cost	NOTES	
Prepare Soil:					
Spread fertilizers, compost	25.10		120.00	4 yards compost, 50 lbs fertilizer	
Rototill	12.55	3.00		1 hr	
Rake, handwork	25.10			2 hrs	
Set drip lines, patch, check	25.10		25.00	2 hrs	$40 drip lines/2 uses plus fittings
Install mulch and anchor	12.55		28.00	1 hr	$200 weed mat/10 yds, anchors
Tighten greenhouse, other	25.10			2 hrs	
Heat, vent, alarm ready	25.10			2 hrs	
Other					
Seed/Transplant:					
Cost of transplants			243.00	450 plants needed/greenhouse, $0.54/3.5" pot	
Transplanting labor	50.20			4 hrs	
Cultivation:					
Drop strings	25.10		5.00	2 hrs	
Clip strings	25.10			2 hrs	
Prune and trellis 7x	329.44			Average: .75 hr/row, 3.75 hrs each time = 26.25 hrs total	
Weed holes, edges 3x	75.30			6 hrs total	
Prune leaves, sweep up 3x	112.95			9 hrs total	
Top plants 9/1	37.65			3 hrs total	
Roll up and down sides	58.99			4 mins/time x 70 days = 4.7 hrs	
Pre-harvest subtotal:	865.33	3.00	421.00	= 1289.33 Pre-harvest cost	

Harvest:

Total yield for greenhouse = 300 15-lb boxes — at 10 lbs marketable fruit/plant

Total hours to harvest greenhouse 60 hrs — average: five 15-lb boxes/hr

	Labor cost	Machinery cost	Product cost	NOTES	
Field to pack house	753.00			at $12.55/hr	60 hrs
Pack house to dock	376.50			at $12.55/hr	at 10 boxes/hr sorting and folding up boxes
Bags, boxes, labels			321.00	$1.00/box, $0.07/label	300 at $1.07
Delivery	30.12	9.60		See Worksheet 1.	
Post Harvest:					
Detrellis and remove plants	75.30			6 hrs total	
Sweep and fold mulch	12.55			1 hr	
Move drip lines	12.55			1 hr	
Post-harvest subtotal:	2125.35	12.60	742.00	= 2879.95 Harvested cost for greenhouse	
Marketing Costs:					
Labor: sales calls for season (for this crop only)	25.10			Average 10 mins/week for 12 weeks = 2 hrs	
Commissions				Commissions, if any, to growers' co-op, broker, or salesperson	
Farmers' market expense	60.24	4.70	9.00	See Worksheet 1.	
Total Crop Costs:	2210.69	17.30	751.00	= 2978.99 Total crop costs	

Greenhouse & Overhead Costs: 3227.00 Greenhouse annual expenses: $830; greenhouse overhead allocation: $2397. See Worksheet 3.

Total Costs:
Crop & Overhead Total: 6205.99

Sales:	# of units	Price per unit	Total $
Retail:	100.00	48.75	4875.00
Wholesale:	200.00	36.00	7200.00
Other:			0.00
Total units	300.00		
Total Sales:			12075.00

Net Profit: Total sales – total costs =	5869.01	**Net profit for Greenhouse**
Net Profit/Acre:		Not applicable
Cost/Unit:	20.69	Total cost/total units
Net Profit/Unit:	19.56	Net profit/total units

NOTES:

Index

Q

R

S